Cambridge Elements ☰

Elements of Flexible and Large-Area Electronics
edited by
Ravinder Dahiya
University of Glasgow
Luigi Occhipinti
University of Cambridge

A FLEXIBLE MULTI-FUNCTIONAL TOUCH PANEL FOR MULTI-DIMENSIONAL SENSING IN INTERACTIVE DISPLAYS

Shuo Gao
Beihang University
Arokia Nathan
University of Cambridge

CAMBRIDGE
UNIVERSITY PRESS

CAMBRIDGE
UNIVERSITY PRESS

University Printing House, Cambridge CB2 8BS, United Kingdom

One Liberty Plaza, 20th Floor, New York, NY 10006, USA

477 Williamstown Road, Port Melbourne, VIC 3207, Australia

314–321, 3rd Floor, Plot 3, Splendor Forum, Jasola District Centre,
New Delhi – 110025, India

79 Anson Road, #06–04/06, Singapore 079906

Cambridge University Press is part of the University of Cambridge.

It furthers the University's mission by disseminating knowledge in the pursuit of
education, learning, and research at the highest international levels of excellence.

www.cambridge.org
Information on this title: www.cambridge.org/9781108735315
DOI: 10.1017/9781108686532

First published 2019

A catalogue record for this publication is available from the British Library.

ISBN 978-1-108-73531-5 Paperback
ISSN 2398-4015 (online)
ISSN 2514-3840 (print)

Cambridge University Press has no responsibility for the persistence or accuracy of
URLs for external or third-party internet websites referred to in this publication
and does not guarantee that any content on such websites is, or will remain,
accurate or appropriate.

A Flexible Multi-Functional Touch Panel for Multi-Dimensional Sensing in Interactive Displays

Elements of Flexible and Large-Area Electronics

DOI: 10.1017/9781108686532
First published online: June 2019

Shuo Gao
Beihang University

Arokia Nathan
University of Cambridge*

Author for correspondence: Shuo Gao: shuo_gao@buaa.edu.cn; Arokia Nathan: an299@cam.ac.uk

Abstract: Touch panels (TPs) have become an integral part of modern-day lifestyle. To enhance user experience, attributes such as form-factor flexibility, multi-dimensional sensing, low power consumption, and low cost have become highly desirable. This Element addresses the design of multi-functional TPs with integrated concurrent capture of ubiquitous capacitive touch signals and force information. It compares and contrasts interactive technologies and presents design considerations for multi-dimensional touch panels with high detection sensitivity, accuracy, and resolution.

Keywords: capacitive sensing; energy harvesting; flexible form-factor touch panel; force sensing; interactive displays; multi-dimensional sensing, piezoelectric materials

ISBNs: 9781108735315 (PB), 9781108686532 (OC)
ISSNs: 2398-4015 (online), 2514-3840 (print)

* Arokia Nathan is now at Cambridge Touch Technologies, Cambridge.

Contents

1 Human–Machine Interaction–Related Technologies in Interactive Displays

Visual display of information is greatly required in today's highly digital world and constitutes a powerful means of conveying complex information, which cannot conveniently be described in text. Visualizing information is based on the ability of the human eye and brain to perceive and process vast quantities of data in parallel. Its history can be traced to the ancient era, when our ancestors carved images on cave walls and monuments (around 30,000 BC [1]). Mosaic art forms emerged in the third millennium BC [2], with small pieces of glass, stone, or other materials used in combination to display information. These pieces are a counterpart to the pixels in modern electronic display. The electronic display has become the primary human–machine interface in most applications, ranging from mobile phones, tablets, laptops, and desktops to televisions, signage, and domestic electrical appliances, not to mention industrial and analytical equipment.

User interaction with the electronic display has progressed significantly. Through sophisticated hand gestures [3]–[11], the display has evolved to become a highly efficient information exchange device. While interactive displays are currently very popular in mobile electronic devices such as smart phones and tablets, the development of large-area, flexible electronics offers great opportunities for interactive technologies on an even larger scale. Indeed, technologies that were once considered science fiction are now becoming a reality, the transparent display and associated smart surface being a case in point. These technologically significant developments beg the question, "What will be the development trend of interactive technology?" This section reviews current mainstream interactivity techniques and predicts what we believe will be future interactive technologies.

Human–machine interactivity can be categorized based on touch or touch-free gestures. The former is employed primarily in the small- and medium-scale panels used in smart phones and tablets, while the latter is more popular in larger displays [12]. Various techniques for interactivity have been developed. Currently these are based mainly on resistive, capacitive, surface acoustic wave, acoustic pulse recognition, and infrared schemes [3]. Recently, touch-free (e.g., gesture recognition by optical imaging) and force-touch technologies have emerged and are now in commercial devices. These advanced features bring human–machine interactivity to a new level of user experience.

1.1 Touch Interactivity Architectures

A variety of techniques have been proposed and implemented for touch panels, including resistive-, capacitive-, acoustic-, and infrared-based architectures.

Among them, the capacitive-based touch panel has dominated the market in recent years, in view of its excellent optical clarity, high accuracy, and multi-touch support. Other techniques have obvious drawbacks that limit their successful use in commercial interactive displays. For example, resistive-based techniques cannot support multi-touch events; acoustic-based techniques have non-touch areas in the presence of solid contaminants; and infrared-based touch interfaces are prone to low detection accuracy in the presence of strong sunlight. We first review all of the foregoing techniques along with a comparison table that summarizes their merits and drawbacks.

The first generation of touch panels employed resistive-based architectures [4]–[8], in which two transparent electrically resistive layers are separated by spacer dots and connected to conductive bars in the horizontal (x-axis) and the vertical (y-axis) sides, respectively. A voltage applied on one layer can be sensed by the other layer, and vice versa. When the user touches the panel, the two layers are connected at the touch point and work as voltage dividers, and the touch location is then calculated. These first-generation devices were limited to locating a single point, restricting their use for complex gestures.

In capacitive-based touch panels, electrodes are arranged as rows and columns and are separated by an insulating material such as glass or thin film dielectric. When a conductive object comes in contact with the panel surface, the electric field is perturbed, hence changing the capacitance between electrodes [9]–[15]. Capacitive touch panels are most commonly used in smart phones because they support multi-touch sensing without altering the visibility and transparency of the display.

In surface acoustic wave and acoustic pulse recognition interactivity schemes, the touch position is detected by acoustic waves [16]–[21]. In the former, ultrasonic waves are transmitted and reflected in the x- and y-directions. By measuring the touch-induced absorption of the waves, the location can be determined. In acoustic pulse recognition, transducers are fitted at the edges of the touch panel. A touch action on the panel surface generates a sound wave that is then detected by the transducers, digitalized, and subsequently processed to determine the touch position.

In the infrared-based architecture, two adjacent sides of a touch panel are equipped with light-emitting diodes, which face photodetectors on the opposite sides, forming an infrared grid pattern [22]–[26]. The touch object (e.g., finger or stylus) disrupts the grid pattern, from which the touch location is determined.

The techniques described in the preceding paragraphs detect two-dimensional single- or multi-touch, i.e., touch locations on an x–y plane. Table 1.1 summarizes their main pros and cons. Commercial products recently released by Apple support force sensing, expanding touch interactivity to three dimensions

Table 1.1 Comparison of mainstream contact touch technologies

Working Principle	Multi-touch?	Glove Touch?	Hover Touch	Optical Clarity	Outdoor Operability
Resistive	No	Yes	No	Medium	Good
Capacitive	Yes	No	Yes (but very short distance)	Good	No in rain
Acoustic	No	Yes	No	Good	Good
Infrared	Yes	Yes	No	Excellent	No under strong sunlight

[27]–[29]. Here, panel deflection, and hence the corresponding change in capacitance, serves as a measure of the extent of applied force, which is then augmented with a haptic response.

1.2 Touch-Free Interactivity Architecture

While a variety of touch technologies are currently in use in products, touch-free gesture recognition has emerged recently. One current technique relies on locating discrete infrared sources and detectors at different positions on the display edges to construct the touch event. However, imaging is not possible because of the discrete nature of the sensors. The pixelated approach reported recently employs an image sensor integrated at every display pixel, thereby making it possible for the display to view the underlying gestures of the user. Alternately, the event can be remotely triggered by a light pen [30]–[35]. The interactive display can be transparent using, e.g., oxide semiconductor technology, and be able to carry out invisible image capture. This development has the potential for a high technological impact in human interfaces.

Voice recognition is another technique for remote interactivity [36]–[40]. Tremendous progress has been made in this area, with very impressive results. Existing commercial products include Siri and Echo from Apple and Amazon respectively. Even so, challenges remain in voice signal processing and machine limitations of speech perception, particularly with differently spelled but similar-sounding words, and signal recognition in a noisy acoustic background. These problems can be eventually overcome with the use of much faster processers and more memory to bring into consideration contextual information.

1.3 Future Human–Machine Interface

Current mainstream human–machine interactive (HMI) technology–touch panel (TP) has two shortcomings. First, recently developed TPs employ a single sensing technique to detect one certain type of physical signal (one-dimensional sensing), as explained earlier and illustrated in Fig. 1.1a–f. Thus, multiple discrete devices with different sensing capabilities must be embedded into a single system to allow multi-dimensional sensing. For example, optical, temperature, and force sensors are integrated into commercial mobile phones to provide multi-dimensional signal detection functions for customers. However, this results in increased component costs, circuitry complexity, and power consumption. Second, although the energy cost is very small for the individual touch sensors in a TP, their total energy consumption is huge, considering numerous touch panels are intensively used worldwide. Besides optimizing the product design to reduce power consumption,

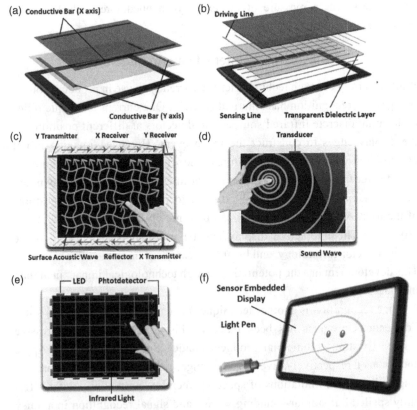

Figure 1.1 Interactivity based on (a) resistive, (b) capacitive, (c) surface acoustic wave, (d) acoustic pulse recognition, and (e) infrared touch architectures, and (f) touch-free interactive display based on image sensor.

which approaches the limits of current technology, harvesting the environmental energy is essential to enhance the lifetime of the battery.

Thus, in the foreseeable future, the key task is to design and implement a multi-functional TP for multi-dimensional sensing along with possible applications to energy harvesting. First, multi-dimensional signals must be detected concurrently, providing customers with a user experience similar to that afforded using multiple mono-dimensional sensors. Second, because TP is a highly commercialized product, the proposed technique should fit well with existing TP techniques to avoid or reduce changes to production lines. Third, potential issues of the proposed technique need to be analysed and addressed. Last but not least, flexibility is a very important attribute for TPs, with the potential to enable and enhance a variety of applications to bring customers novel and advanced experiences.

To achieve these objectives, first, flexible functional materials are expected to be employed, owing to their inherent capabilities for flexibility and response to external stimuli. Specifically, in line with this work, piezoelectric materials will be used to assemble a prototype for demonstrating the authors' strategy, because of their intrinsic ability to convert mechanical stress to electric charges, providing the functions of force touch detection and energy harvesting. Second, the piezoelectric materials will be combined with capacitive touch panels, which dominate the TP market [41]–[43]. Third, algorithms on how to interpret these two signals will be developed.

1.4 Outline of this Element

This Element charts the authors' work on the understanding of capacitive TP and piezoelectric materials and the development of a multi-functional touch panel from theoretical analysis to touch panel fabrication and algorithm design. Section 2 provides literature reviews on capacitive touch panel and piezoelectric materials. The multi-functional touch panel for concurrently sensing force and capacitive stimuli is proposed at the end of this section. In Section 3, a theoretical analysis of the proposed technique is provided in mechanical and electrical terms, followed by a description of preliminary experiments for the purpose of validating the concept. The proposed multi-functional touch panel is fabricated and measured in Section 4. The experimental results demonstrate its good mechanical and electric response to touch events. Section 5 focuses on design and implementation of the algorithm for interpreting the force touch signal. Two practical issues facing force touch sensing are first addressed with the help of the capacitive touch signal: static force touch detection and stress propagation. Next, an algorithm is developed to achieve concurrent force touch detection and energy harvesting. Finally, the conclusions and technological outlook are described in Section 6.

2 Reviews on Capacitive Touch Panel– and Piezoelectric–Related Technologies

In the previous section, the need for a simple-structured multi-functional touch panel was explained, along with the design requirements for the multi-functional touch panel. To develop a piezoelectric material–based capacitive touch panel, it is necessary to gain an understanding of the capacitive touch panel and piezoelectric materials.

In this section, brief literature reviews are provided, first in terms of the working principles of projected capacitive touch panels and piezoelectric materials. Next, the design considerations of embedding piezoelectric material into capacitive touch panels are given through theoretical analysis and practical experiments. Finally, a multi-layered stack-up is proposed to achieve multi-functionality.

2.1 A Brief Overview of the Projected Capacitance Touch Panel

2.1.1 Working Principle and Panel Architecture

A projected capacitance touch panel system detects capacitance variations at electrodes to recognize the touch event [4], [9]. When a conductive object (e.g., human finger) is in close proximity to or in contact with the touch panel, the surrounding electromagnetic field is perturbed, altering the capacitance between electrodes. This is sensed and the signal is then digitalized and sent to the microcontroller (MCU) to determine the potential touch event and corresponding location. Two architectures broadly used are based on modulation of the self-capacitance and mutual capacitance.

In self-capacitance TP, the capacitance between electrodes to ground is measured [9], [44]–[46]. When a conductive object is approaching the electrode (electrodes are normally protected by a layer of dielectric cover, e.g., glass, and therefore cannot be contacted), the capacitance from the electrode to the ground is increased, and hence a touch event is detected. Two types of self-capacitance are constructed – multi-pad and row-and-column [9] – as shown in Fig. 2.1. In a multi-pad structure, each pad is connected with the controller individually; thus multi-touch is supported. In a row-and-column structure, each of the rows and columns is an electrode, instead of a pad as in a multi-pad structure, and individually connected with the processor. Although each intersection of rows and columns indicates a unique location on the touch panel, it cannot support multi-touch sensing because each electrode is measured, instead of each

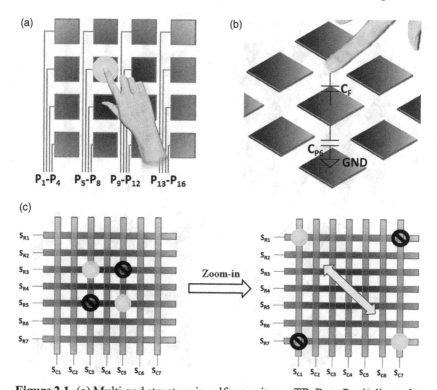

Figure 2.1 (a) Multi-pad structure in self-capacitance TP; P_1 to P_{16} indicate the number of the touch pads. The yellow point represents the touch location. (b) Working principle of multi-pad structured self-capacitance TP. CP_6 is the capacitance between touch-pad P_6 to ground and C_F is the finger touch induced capacitance. (c) Ghost points in row-and-column structured self-capacitance. SR_1 to SR_7 and SC_1 to SC_7 indicate the row and column sensing electrode 1 to 7, respectively. The yellow points and black block signs are the real touch locations and ghost point locations.

intersection. Thus when multi-touch is performed, ghost points are made as illustrated in Fig. 2.1c.

However, the zoom-in/zoom-out function still works, as the distances between the interpreted touch locations are calculated by software. When the distance increases, a zoom-in action can be interpreted. In contrast, a decrement of distance between registered touch locations indicates a zoom-out action. One advantage of the self-capacitance structure is its ability to detect hover touch and glove touch, as long-distance field projection is normally used [46].

Alternatively, in mutual-capacitance TP, the mutual capacitance between two electrodes is measured [46], [47]. In mutual-capacitance–based techniques, row-and-column structured electrodes are normally employed [47].

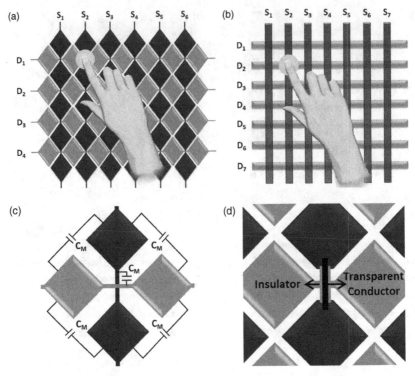

Figure 2.2 (a), (b) Mutual-capacitance diamond structure and bar structure. The yellow points indicate the touch locations. (c), (d) Working principle of mutual-capacitance structure.

Electrodes in rows function as driving lines and electrodes in columns work as sensing lines, or vice versa. Each intersection of rows and columns indicates a unique location (each location can be treated as a pixel; hence many image-related techniques are used to process touch signals), and each intersection will be sensed individually. By periodically scanning electrode intersections, multi-touch detection is supported. As shown in Fig. 2.2, electrodes in rows are arranged from D_0 to D_N, which are powered separately. The capacitance values of intersections with the sensing lines from S_1 to S_M will be detected in sequence to achieve multi-touch detection. When a conductive object contacts the panel cover, the mutual capacitance is decreased because charges are taken by the human finger, as conceptually shown in Fig. 2.3. Compared to the self-capacitance architecture TP, one major disadvantage of mutual-capacitance TP is considerable scanning time for a full-panel measurement. State-of-the-art commercial products have a sensing rate from 20 Hz to 200 Hz [9], while some laboratory-used and -developed touch panels can achieve a higher sensing rate, up to 6,400 Hz[3], [14], [46]–[54].

Table 2.1 Main distinctions between self-capacitance and mutual-capacitance structures

Characteristic	Self-Capacitance	Mutual- Capacitance
Electrode type	Sensing	Driving and sensing
Number of layers	1 or 2	1 or 2
Electrode design	Multi-pad/row-and–column	Unique electrode intersections
Scanning method	Each electrode	Each electrode intersection
Whole panel scanning time	⊠	⊠ ⊠
Measured capacitance	Capacitance of electrode to ground	Capacitance between electrodes
Ghost point	Yes for row-and-column structure	No
EMI robustness	Bad	Good

EMI, electro-magnetic interference.
Modified from [2]–[6].

The main distinctions between self-capacitance and mutual-capacitance structures are summarized in Table 2.1.

2.1.2 Touch Panel Construction

Almost all the projected capacitive touch panels share two basic features in their construction [46]. First, the touch surface is above the sensing circuits; second, all the components are fixed, which means there are no moving parts. A typical two-layer projected capacitance construction concept is shown in Fig. 2.4. Two transparent thin-film ITO conductors are separated by a thin-film insulator (normally glass or polyethylene terephthalate [PET]), and a touch surface is set on top of them.

The sheet resistance and line widths of the patterned indium tin oxide (ITO) layer are normally 150 $\Omega/\curlyvee$ and 20 μm when glass is used as substrate [9]. In contrast, when PET is employed as a substrate, the line widths are typically 100 to 200 μm [9], due to the reduced flatness compared to glass. For glass substrate–related ITO patterning, photolithographic methods are widely used. As to the PET substrate, more techniques can be applied for ITO patterning, such as panel printing [9]. Although the sheet resistance and line widths of the PET substrate–based patterning are higher and larger than those of the glass substrate–based patterning, the advantage of using a PET substrate is its thinness. The thickness of a PET substrate is usually 50 μm to 100 μm [9], [44], [55], [56]. Alternatively, the thickness of a glass substrate is 0.2 mm to 0.4 mm [9], [44]. A detailed comparison of a PET substrate versus glass substrate is given in Table 2.2. Optical clearance adhesive (OCA) is widely used to glue the multi-layered structure [9], [46].

Table 2.2 Main distinctions between self-capacitance and mutual-capacitance structures

Characteristic	PET	Glass
Glass transition Temperature	70°C	570°C
Aging effects	Yellowing, curling, surface deformation	No known effect
Transparency	85%	≥90%
Resolution	10–30 μm	1 μm
Stack-up	Thinner	Thicker
Weight	Lighter	Heavier
Lamination yield	Excellent	Good
Cost	$$ (was less than for glass)	$

Modified from [2], [3].

(a)

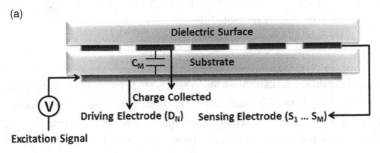

(b)

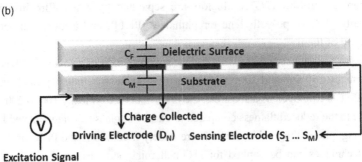

Figure 2.3 Working principle of mutual-capacitance structure (a) without and (b) with a finger touch.

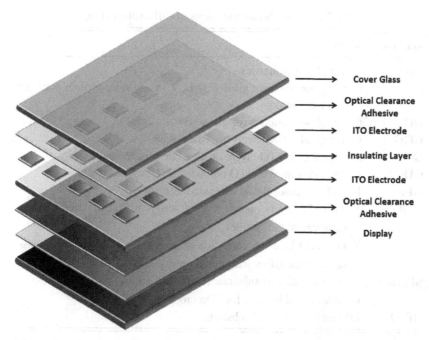

	Cover Glass
	Optical Clearance Adhesive
	ITO Electrode
	Insulating Layer
	ITO Electrode
	Optical Clearance Adhesive
	Display

Figure 2.4 Structure of a typical two-layer projected capacitive touch panel.

Besides the stack-up shown in Fig. 2.3, there are many other stack-ups widely used in industry. The reasons touch-module makers select one but not others are based on considerations such as transmissivity, thickness, weight, and cost. The symbols and meanings for different stack-ups are summarized in Table 2.3.

2.1.3 Capacitance Measurement Methods

Capacitance can be measured by a variety of methods. In general, touch-induced capacitance can be detected by measuring the change of RC constant, impedance, and amount of transferred charges. The main methods are relaxation oscillator [57], charge time versus voltage [58], voltage divider [59], and charge transfer [60].

When a charge time versus voltage-based method is employed, a power supply is used to charge the capacitor (*C*). At the meantime, a resistor with known resistance (*R*) is connected to the capacitor, allowing it to discharge. The voltage across the capacitor (*V*) can be expressed as

$$V = V_0 e^{-\frac{t}{RC}},\tag{2.1}$$

Table 2.3 Symbols and meanings of different stack-ups

Symbol	Meaning
(G)	Cover glass (or plastic)
G	Cover glass or sensor glass with ITO on one side, or plain glass for film lamination
GG	Cover glass + one sensor glass
GGG	Cover glass + two sheets of sensor glass
G#	# = Number of ITO layers on one side of sensor glass
GIF	F = Sensor film with ITO on one side, laminated to glass
GFF	FF = Two sensor films, laminated to glass
GF#	1. Two ITP layers on one side of sensor film, laminated to glass (also called GF-Single)
	2. One ITO layer on each side of sensor film, laminated to glass (also called GFxy with metal mesh)
SITO	ITO on one side of substrate (single-sided); usually includes metal bridges for Y to cross X
DITO	ITO on both sides of substrate

Modified from [6].

where t indicates time and V_0 the initial voltage at $t = 0$. By detecting V, the capacitance can be interpreted.

In the voltage divider method, a capacitor with known value is connected with a (sensing) capacitor for measurement. As the two capacitors are in series, they contain the same amount of charges. However, the voltages across them are different, depending on their capacitance values. Through the relationship between the voltage, charge, and capacitance ($Q = CV$), the unknown capacitance can be interpreted.

In transferred charge–based methods, a current source (or a voltage source) is employed to provide a stable periodic signal [61]. As shown in the black components in Fig. 2.5a, by integrating the charges accumulated on the electrode capacitor ($C_{\text{Electrode}}$), the charge amplifier, consisting of an operational amplifier and feedback components (feedback resistor R_F and feedback capacitor C_F) outputs a voltage signal (V_{Out}). The amplitude of the output is positively proportional to the ratio of the $C_{\text{Electrode}}$ to the C_F. When a touch event happens, a touch induced capacitor (C_{Touch}) is paralleled to the original electrode capacitor (C_F), as shown in the grey components in Fig. 2.5a. Thus the input capacitance increases, resulting in a boost in the output voltage. In this way, a touch event is detected.

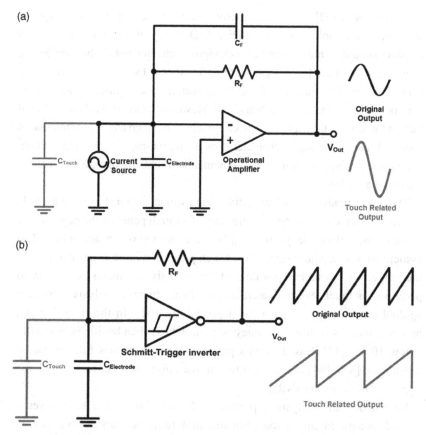

Figure 2.5 Working principles of (a) charge amplifier and (b) relaxation oscillator techniques based touch detection.

As to the relaxation oscillator–based capacitance measurement, a non-linear electronic circuit is used to generate a periodic non-sinusoidal (e.g., triangular wave or square wave) signal. As shown in the black components in Fig. 2.5b, through the feedback resistor (R_F), the input capacitor ($C_{Electrode}$) is charged. When the voltage across the input capacitor $C_{Electrode}$ exceeds a certain threshold, the inverter is triggered, and then the output becomes zero. This process happens periodically, and the period is controlled by the feedback resistor R_F and input capacitor $C_{Electrode}$. When a touch event happens, the touch-induced capacitor C_{Touch} increases the RC constant, hence decreasing the frequency of the output signal, as shown in the grey part of Fig. 2.5b.

2.1.4 Characteristics of Capacitive Touch Signals

To accurately interpret capacitive touch signals in terms of position and presence, and avoid touch misregistrations, digital signal processing (DSP) algorithms [47], [62], [63] are normally applied to the digitized touch signals. To design and implement high-efficiency DSP algorithms, the first thing to consider should be the capacitive touch signal's characteristics. In practice, the characteristics of the touch signals are different case by case. They are highly dependent on the structure of the touch panel, measurement methodology, environmental noise, and user behaviour. However, there are still some shared factors for most of the capacitive touch signals. A detailed analysis of the shared human finger touch signal characteristics of mutual-capacitance architecture, which is intensively used in current commercial mobile phones, is provided in the text that follows.

The property analysis of capacitive touch signals can be based on a single electrode intersection or on a whole scan of a touch panel. If the capacitance value of each electrode intersection is treated as a pixel value, and the data associated with a whole scan of the touch panel is treated as an image, then the property analysis is equivalent to the analysis of a touch event–related pixel and a touch event–related image. Time domain analysis is widely applied on the touch event–related pixel [47], [62]. In the time domain, human touch is a low-frequency signal, whose bandwidth is normally below 10 Hz [47]. Based on this property, many low-pass filtering based techniques [47], [64] are applied to remove noise, in order to improve the signal–to-noise ratio (SNR).

Spatial domain analysis is performed for analysis of the touch event–related image. In the spatial domain, first touch signals are viewed as low spatial frequency signals [63] because the size of the human finger is normally 7 mm to 15 mm, and traditionally, spacing of two adjacent electrode intersections is around 5 mm [65]. When a finger touch is performed, normally 3×3 electrode intersections are affected [29]. As the spacing in recent touch panels is becoming smaller and smaller, more electrode intersections are affected. In an experiment carried out with an 80×80 Blackberry laboratory touch panel, the spacing for the sensing array is only 2 mm; thus more electrode intersections ($\sim 11 \times 11$) are affected as shown in Fig. 2.6. Within the touch event–affected region, the capacitance values of adjacent electrode intersections are similar; hence the touch signal offers a low spatial character. This character is important for smoothing noise spikes.

Second, the capacitance intensities of electrode intersections within the touch-affected area follow a certain distribution (e.g., Gaussian) [29], [65], [66]. This is due to the shape of the human finger. Various subpixel inter-polation algorithms [29], [63] for detecting accurate touch positions are developed based on this characteristic. Some widely used subpixel estimators

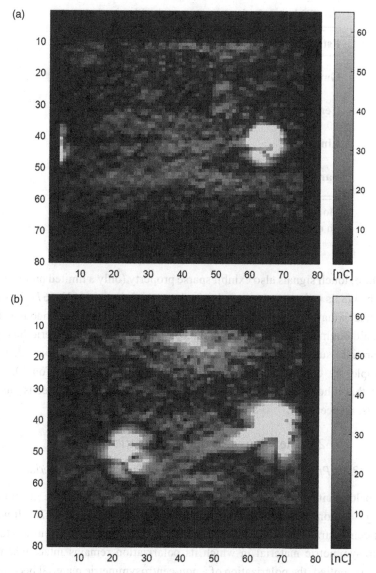

Figure 2.6 (a) Single touch event–related image and (b) multi-touch event–related image.

are summarized in Table 2.4. Furthermore, algorithms for avoiding mistouch registrations are also based on this characteristic. For example, when a user's cheek touches the touch panel, although the capacitance intensity exceeds the touch determination threshold, the cheek touch activity is not registered due to its intensity distribution; hence an important phone call will not be disturbed by a user's unintentional operations.

Table 2.4 Estimators for subpixel interpolation

Estimator	Equation
Gaussian	$\dfrac{\ln r_{i-1} - \ln r_{i+1}}{2(\ln r_{i-1} - \ln r_i + \ln r_{i+1})}$
Centre of mass	$\dfrac{r_{i+1} - r_{i-1}}{r_{i-1} + r_i + r_{i+1}}$
Linear	$\dfrac{r_{i+1} - r_{i-1}}{r_i - min\{r_{i-1}, r_{i+1}\}}$
Parabolic	$\dfrac{r_{i-1} - r_{i+1}}{r_{i-1} - 2r_i + r_{i+1}}$

Modified from [28]. r_i indicates the capacitance value at the ith row.

Third, touch signals also exhibit sparse property (only a limited number of touch signals exist simultaneously) in the spatial domain [63], [67]. Under practical situations, only one or two touch events happen simultaneously. In fact, although a touch panel scans all its electrode intersections periodically, it normally supports only a few touch events during one scan period. For example, by employing a multi-touch testing software (MultiTouch), we learn that iPhone 6S and iPad Pro can support 5 and 17 simultaneous touch events, respectively.

2.2 Brief Overview of Piezoelectric Material

2.2.1 Principle and Characterization of Piezoelectric Material

Piezoelectricity is the phenomenon by which a mechanical force generates charges in non-centrosymmetric structured solid materials [68]–[70]. It was first demonstrated in 1880 by C. Linnaeus and F. Aepinus. In contrast to a centrosymmetric material in which its polarization remains intact when a force is applied, the polarization of a non-centrosymmetric material becomes either positive or negative depending on the direction of the applied force. This is illustrated in Fig. 2.7a and b. The structure of a specific piezoelectric material, PVDF in its β-phase, is shown in Fig. 2.7c to provide a direct view of how the polarization changes due to the structure of the material. Piezoelectric materials also demonstrate the inverse piezoelectric phenomenon, in which a mechanical deformation is induced when an external electric field is applied [71]. The phenomenon of force-induced charge is a designated direct effect, and its counterpart is known as the motor effect [71]. To quantify piezoelectric performance, piezoelectric equations and coefficients

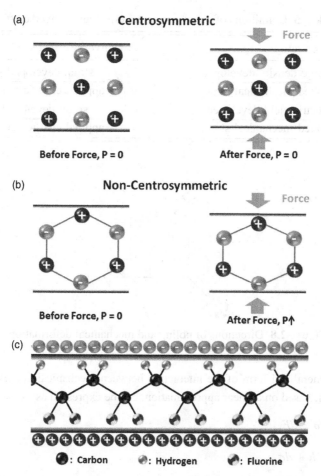

Figure 2.7 (a, b) Change of polarization in centrosymmetric and non-centrosymmetric structures. (c) Structure of PVDF β-phase, modified from [30].

can be used. The piezoelectric equations and coefficients are summarized in the text that follows. The measurement of the main coefficient is discussed in the next section.

The charge and voltage piezoelectric constants are denoted as d and g, respectively. Their relationship is expressed as [72]:

$$d = \varepsilon_0 \varepsilon_r g, \tag{2.2}$$

where ε_0 denotes the vacuum permittivity (8.85 pF/m), and ε_r indicates the relative permittivity (dielectric constant) of the piezoelectric material. Electrically, d and g represent short and open circuit conditions for the piezoelectric materials. In direct and converse effects, d and g are defined as [72] shown in Table 2.5.

Table 2.5 Definitions of d and g in direct effect and converse effect

Direct effect	Converse effect
$d = \dfrac{\text{charge density developed}}{\text{applied mechanical stress}}$	$d = \dfrac{\text{strain developed}}{\text{applied electric field}}$
$g = \dfrac{\text{electric field developed}}{\text{applied mechanical stress}}$	$g = \dfrac{\text{strain developed}}{\text{applied charge density}}$

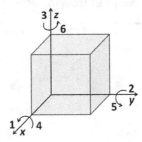

Figure 2.8 Directions of poling and mechanical deformation.

The generalized form of the interaction between mechanical and electrical behaviour, based on a linear approximation, can be expressed as

$$\varepsilon = \varepsilon^E \sigma + dE, \tag{2.3}$$

$$D = \varepsilon^\sigma E + d\sigma, \tag{2.4}$$

where S and T denote the strain and the applied stress, E and D represent the electric field strength and the dielectric displacement, and s and ε indicate the compliance and the permittivity.

As the directions of poling and mechanical deformation can be varied, it is desirable to identify the axes of a sample when specifying the parameters. Normally a model similar to that shown in Fig. 2.8 is used to indicate the directions of poling and mechanical deformation [73]. Here the three axes (x, y, and z) are perpendicular to each other. Of these, the '3-axis' (z) usually represents the poled direction. Shear strains related to directions 1, 2, and 3 are denoted as 4, 5, and 6. From this coordinate system, the generalized form incorporates directional notation. Thus, when a piezoelectric sample is poled in the '3-direction', and stress is also applied in the '3-direction', the relationship between d, D, and T becomes

$$d_{33} = [\delta D_3 / \delta T_3]_E. \tag{2.5}$$

The electrodes are on the faces in the '3-direction' and the external electric field is constant. Eq. (2.5) can be rewritten as

$$d_{33} = (Q/F), \tag{2.6}$$

where Q is the charge developed, and F is the applied force. Equation (2.6) offers a means to measure d_{33} using the direct method. d_{33} can also be measured using the indirect method, which can be explained by rewriting Eq. (2.2) as

$$d_{33} = [\delta S_3 / \delta E_3]_T, \tag{2.7}$$

In this case, the experiment is carried out under constant stress. The direct (Berlincourt) method is introduced in detail in the text that follows, as force touch detection is similar to the inverse procedure of the direct method.

2.2.2 Challenges of Piezoelectric Material–Based Force Touch Detection in Interactive Displays

As explained in the Appendix, the force detection process is similar to the inverse process of measuring the piezoelectric d_{33} coefficient using the Berlincourt method [74]. Here, d_{33} is determined by measuring the force-induced charges. Equations under the condition of zero intensity electric field for this can be summarized as

$$\sigma = F/A, \tag{2.8}$$

$$P_3 = d_{33}\sigma; \tag{2.9}$$

$$Q = AP_3 = d_{33}F, \tag{2.10}$$

where P_3 is the force-induced polarization in the poled direction, and σ indicates stress. As the sample is under compression, scalar expression of Hooke's law is used. As stated in the Appendix, the piezoelectric d_{33} coefficient is not constant [75]–[77], and the actual value depends on many factors. Thus, it is necessary to address how these factors affect the accuracy of force touch detection, when the piezoelectric material is integrated into a touch panel system. Two assumptions are made: first, that the touch panel is designed for interactive displays in a current commercial consumer product (e.g. a mobile phone) and second, that the force touches are performed by human fingers. The following discussion is based on these two assumptions.

The frequency of the force touch is limited to a certain frequency band, and is highly dependent on the individual's behaviour and the type of software application. For example, drawing applications normally don't require fast touch actions, while game applications may demand quick touches. Thus, when the piezoelectric material is used for force detection, the same force can generate different charges when the touch speed differs, decreasing the force touch detection accuracy.

The effects of force amplitude and time of the pre-load can be neglected because after the product is assembled, the amplitude of the pre-load is almost fixed, and will not dramatically change during a short time. The time of the pre-load is normally at the scale of days and months, indicating that the piezoelectric d_{33} coefficient is in the stable region. The geometry of the sample and loading are strongly related to the product design; thus the effect of geometry is not discussed here.

The main environmental factors are the temperature and EMI interference. In practice, the main contributor to these two factors usually is the interactive display itself. The electric power source, touch function, and display function can be the main EMI contributors, affecting both the piezoelectric d_{33} coefficient and the accuracy of the readout circuit. As shown in [62], the noise from the charger and LCD can exceed 1 V, strongly weakening the force touch accuracy. However, these two types of noise can be cancelled using correlated double sampling (CDS), as described in [47]. Another type of noise from the interactive display can originate from other touch sensing functions. Here, only the project capacitive touch sensing is discussed. DC signals are widely used for resistive and optical touch sensing functions, the effects of which can be treated and removed as offset [35], [79]. The capacitive touch function employs a sensing signal of around 100 kHz [64] to detect capacitive touch events, as shown in [62]. Thus, by using low-pass and high-pass filters, the capacitive and force touch signals can be separated.

Two main sources contributing to changes in temperature are the external AC electric field and electrical components of the interactive display. The AC electric field can result in small vibrations of the piezoelectric film, giving rise to a boost in temperature [72]. However, as the EMI-induced heat is normally much smaller than that from the processor of the interactive display, it can be neglected. Temperature changes made by the processor or other components can give rise to charge generation, due to the pyroelectric behaviour of ferroelectric materials. The incremental rate is very slow, indicating that the frequency is much lower than the force touch signal. Hence, the pyroelectric signal can be filtered out or cancelled using CDS-related techniques [47], as it is a time-correlated signal.

Besides the instability issue of the piezoelectric d_{33} coefficient previously discussed, there are two challenges facing force touch detection using piezoelectric materials in interactive displays: static force detection and elimination of force interferences. Two factors affect the static force detection. First, piezoelectric materials normally offer high relative permittivity [78], which results in

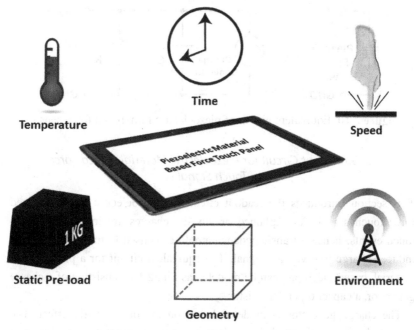

Figure 2.9 Main factors affecting force touch sensing in a piezoelectric material based interactive display.

charge dissipation. Second, thermal drift gives rise to charge generation due to the pyroelectric phenomenon [72]. Elimination of force interferences is complicated by charges induced by adjacent force touches, which can be interpreted by the system as a light force touch, giving rise to force touch misregistration.

The foregoing discussion shows that there are many factors that can influence the accuracy of force touch detection in a piezoelectric material–based interactive display, as shown in Fig. 2.9. Some factors among these can be neglected, such as time, static pre-load, and temperature, as they do not change dramatically over a short period of time. Nevertheless, the touch speed and geometry of the touch object may strongly disturb interpretation of the force signal. For example, the extreme touch speed case is a DC force touch versus a high-speed touch, because although the same force is used, the output signals are different. As to the case of various touch object geometries, different amounts of stress can be induced by the same force when the contact areas are different. As the touch detection accuracy can be undermined by the factors explained earlier, the readout circuit design is vital for achieving high detection accuracy. The following section discusses readout circuit designs for piezoelectric material–based human force touch sensing.

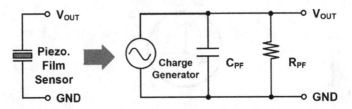

Figure 2.10 Equivalent circuit of piezoelectric film–based force sensor.

2.2.3 Readout Circuit for Piezoelectric Material–Based Force Touch Signal

This section introduces the readout circuit for piezoelectric material–based force touch signal. As explained previously, charges are induced by force touch events; hence a transimpedance amplifier is used to transfer the force-induced charge to a voltage signal. The equivalent circuit for a piezoelectric material–based touch pad can be modeled as Fig. 2.10, consisting of a charge generator, a capacitor, and a resistor.

The charge generator is modeled based on the piezoelectric effect. The magnitude of the generated charge depends on the piezoelectric coefficient of the material and the applied force. The capacitance of the touch pad can be calculated by the following equation:

$$C = \frac{A}{d} e_0 e_r, \tag{2.11}$$

where A is the overlap area of the electrode, d the piezoelectric film's thickness, and e_0 and e_r the vacuum permittivity and relative permittivity, respectively. The resistance of the piezoelectric film touch pad is expressed as

$$R = \rho \frac{l}{A}, \tag{2.12}$$

where ρ denotes the resistivity of the piezoelectric material. The resistivity of the piezoelectric material is generally very large. For example, the electrical resistivity of PVDF [80] is around 2×10^{14} $\Omega \cdot$cm and it drops with increasing frequency, giving rise to a drop in the piezo film resistance (R_{PF}). Considering that the touch events take place only at low frequencies and due to the huge resistivity of piezoelectric material, R_{PF} is broadly treated as open circuit and omitted in low-frequency analysis [82], [83]. Hence the Eq. 2.12 is not described as a function of frequency.

Figure 2.11 illustrates the transimpedance amplifier readout circuit. Here, C_F and R_F are the feedback capacitor and resistor. The main parameters of the circuit are described as

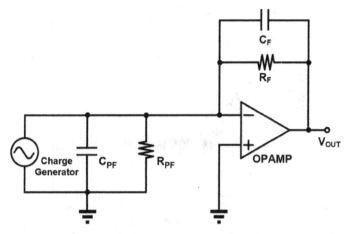

Figure 2.11 Piezoelectric sensor and TIA-based readout circuit.

$$V_{OUT} = -\frac{Q}{C_F} \tag{2.13}$$

$$\tau = R_F C_F \tag{2.14}$$

$$f_{cutoff} = \frac{1}{2\pi R_F C_F} \tag{2.15}$$

From Eq. (2.13) to (2.15), we can conclude that: the circuit gain does not depend on the touch pad's capacitance; however, it needs to be taken into consideration for noise analysis (detailed in the next section). A smaller feedback capacitor gives rise to higher gain, but resulting in a high cutoff frequency as well. Hence, to acquire the low-frequency signal down to few hertz, a high R_F (e.g. 1 GΩ) is used.

Another important design consideration is the bandwidth of the force-induced electric signal. The signal frequency depends on the speed of the touch action. The signal frequency is correlated to the speed a user performs the force touch. For example, the frequency of human touch action may be limited to within 10 Hz (i.e. a maximum number of 10 taps in 1 second). However, the force touch signals from an oscilloscope, as shown in Fig. 2.12, show that the property of the force touch signal in the frequency domain can stretch up to several kilohertz.

2.3 Proposed Multi-functional Touch Panel

Based on the aforementioned literature reviews on capacitive touch panels and piezoelectric materials, the simple-structured multi-layered multi-functional stack-up shown in Fig. 2.13 is proposed for multi-dimensional sensing and

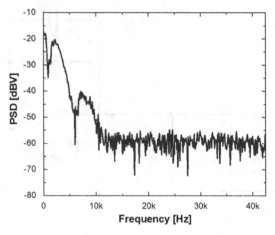

Figure 2.12 Power spectrum density of force touch signals.

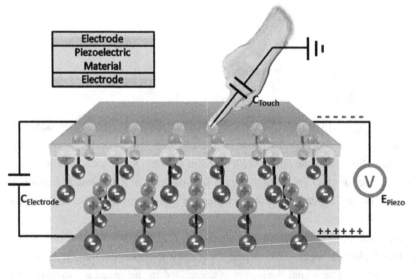

Figure 2.13 Proposed multi-functional stack-up.

energy harvesting in an interactive display. The piezoelectric material functions as an insulating, force touch sensing, and energy harvesting layer. A theoretical analysis of both electrical and mechanical aspects of the proposed technique is provided in the following section.

2.4 Conclusion

This section provides a brief literature review on capacitive touch panels and piezoelectric materials. With an understanding of these two techniques, a multi-layered stack-up is proposed, which is expected to be able to detect capacitive touch and convert mechanical stress to electrical energy. The content in this chapter helps readers to gain a deep insight on both capacitive touch panel–related techniques and working principles of piezoelectric materials, inspiring readers to smartly integrate these two techniques for novel applications.

3 Mechanical and Electrical Analysis of Interactive Stack-ups

The previous section proposed and briefly discussed an approach to implement multi-functionality in capacitance TPs by utilizing force-voltage responsivity of piezoelectric materials, which are broadly used for force sensing [68], [69], [84]–[91]. With the piezoelectric material–related technique, polarization shift induced charges are generated on the surface and transferred by electrodes when a force load is applied. The amplitude of the electric signal produced is positively linear correlated to the strength of the force. Therefore, when the piezoelectric material is used as the dielectric layer for the capacitive touch sensors, the induced charge can be collected by the touch sensors for force interpretation and energy harvesting, depending on different purposes.

However, as explained in Section 1, the proposed technique should provide the user with at least a similar or advanced user experience. Evaluation of the attributes presented here in comparison with those in state-of-the-art commercially available devices is required. The conventional capacitive touch detection technology has no energy-harvesting function to date, while in the scheme presented here, force-induced charge generation can lead to energy savings. In the iPhone X, an additional piezoresistive layer is integrated above the backlight of the display to detect resistance changes with applied force. However, in previous iPhone models force touch is sensed through capacitance change on deflection; the obvious drawback here is the insensitivity of the force touch at boundaries because of the negligible resulting displacement, which serves to decrease the detection accuracy. Thus, in this section, force detection performances, in terms of responsivity and sensitivity, of the piezoelectric material–based touch panels are examined, to learn whether or not the proposed technique can potentially provide advanced force touch detection experience.

Four stack-up architectures intensively adopted in industry are studied and conceptually drawn in Fig. 3.1. A piezoelectric thin film layer (~20 μm) is underneath the touch panel's cover glass (~0.5 mm). The electrodes' thicknesses normally range from tens to hundreds of nanometres, which is much thinner compared to the piezoelectric force sensing layer, PET substrate layer,

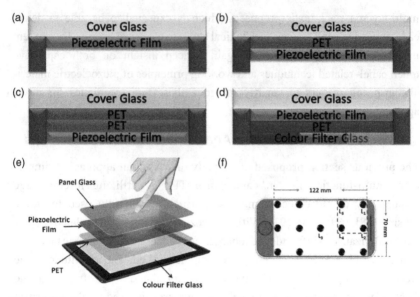

Figure 3.1 Four proposed stack-ups of touch panels (a– d). The electrodes are on and underneath the piezoelectric film layer. (e) Conceptual force touch event applied on one of the stack-ups. (f) Top view of the investigated touch locations.

and cover glass; hence they are not depicted in the figures. As shown in Fig. 3.1e, when a force touch event occurs on the touch panel surface, the stress transferred to the piezoelectric film layer will give rise to the induction of charges, which will be collected and calculated to interpret the force level. The mechanical and electrical properties of the touch panel strongly affect the force touch sensitivity. In the text that follows, studies on these two properties are provided, based on theoretical analysis and simulation results.

After acquiring the force-induced electric signal, a challenge is to success-fully interpret the capacitive touch signal and force touch signal. At the end of the chapter, the separation method for capacitive touch and force touch signals is provided, with experimental demonstration.

3.1 Mechanical Analysis of Proposed Touch Panels

In this section, analysis of the mechanical response of the touch panel to force touches is given. Here, the relationship among force, stress, strain, panel thick-ness, and displacement is investigated. The mechanical properties for different piezoelectric film thicknesses are investigated. The parameters of the proposed stack-ups and properties of the materials are illustrated in Table 3.1 and Table 3.2, respectively. Electrode layers are not given, as they are very thin compared to the other layers; hence, electrodes will not affect the mechanical investigation. The

Table 3.1 Dimension parameters of the proposed stack-ups

	Thickness (m)	Length (mm)	Width (mm)
Panel glass	0.5×10^{-3}	122	70
Piezo. film	$(10, 25, 50, 100) \times 10^{-6}$	122	70
PET	50×10^{-6}	122	70
Colour Filter Glass	0.4×10^{-3}	122	70

Table 3.2 Mechanical and piezoelectrical properties of the proposed stack-ups

	Young's modulus (Pa)	Poisson Ratio	Density (kg/m^3)	Piezo. coeff. (nC/N)
Panel glass	7.4×10^{10}	0.3	2200	———
Piezo. Film	8.3×10^{9}	0.18	1780	38
PET	3.1×10^{9}	0.37	1380	———
Colour filter glass	7.5×10^{9}	0.25	2500	———

From [28], [82], [92]–[98].

touch panels can be assumed as thin plates as their thicknesses are far smaller than the widths and lengths. To investigate the mechanical response of the proposed stack-ups, more investigation points would yield higher accuracy. However, we cannot simulate infinite contact points in practice; hence we select some representative points for the study. Here we evenly divide the respective edges into 4 and 3, obtaining 12 locations. The studied points are mainly at the corner or adjacent to the frames, where insensitivities arise in capacitive-based force sensing. The centre point of the touch panel is also investigated because it is the most sensitive sensing point in capacitive-based force sensing. Second, in the Kirchhoff–Love plate theory based on the simply supported model, the stress and displacement are given only for the centre point. Therefore, this point to evaluate the accuracy of the model that we build in the COMSOL simulation environment. Owing to the symmetric property of the touch locations, only five of them are analyzed, as depicted in Fig. 3.1f.

The force is assumed to be uniformly distributed over the contact area. The boundary conditions of the touch panel lie between a simple supported case and a fully clamped case. For the latter, there is little literature available because of its high complexity. Thus, numerical results from finite element simulation software are used for analysis. In this section, only the simply supported case is investigated. Furthermore, as the thickness and Young's modulus of the glass panel are much larger than those of other layers, the glass panel dominates in the mechanical analysis. As a result, the other layers are neglected.

Navier–Stokes double Fourier series solution [15] is used to obtain the displacement (Eq. 3.1) at a particular position (x, y) from a point force applied at a particular location (ζ, η). Some of the assumptions of Kirchhoff–Love plate theory are as follows:

1. The thickness of the plate is much smaller than all the other physical dimensions.
2. The displacements of the plate are small compared to the plate thickness.
3. The material is linear elastic.
4. Plane strains are small compared to unity.

The touch panel's width and length are of centimetre scale. The thickness is around 1 mm. The displacement of the plate is in micrometre scale, which is much smaller than the thickness. For practical force touch and working temperature conditions, the materials constructed in the touch panel can be assumed as linear elastic [72]. (When the piezoelectric material works in a high temperature or extremely strong force regime, the structure of the film changes; hence the assumption of linear elasticity will not hold, resulting in a non-linear relationship between the force and generated charges, giving rise to an inaccurate force interpretation. In this research, the touch panel equipped in interactive displays works in a relatively stable temperature environment, and the force amplitude is within an acceptable range [several newtons], and hence the assumption holds.) Because of the high Young's modulus of the materials, the plane strains are much smaller compared to unity. Therefore, our model meets all of these criteria. The closed-form solution is expressed as

$$w(x,y,\zeta,\eta) = \frac{\displaystyle\sum_{n=1}^{\infty}\sum_{m=1}^{\infty}\frac{4}{\pi^4 abD}\sin\left(\frac{n\pi x}{a}\right)\sin\left(\frac{m\pi y}{b}\right)\sin\left(\frac{n\pi\zeta}{a}\right)\sin\left(\frac{m\pi\eta}{b}\right)}{\left(\frac{n^2}{a^2}+\frac{m^2}{b^2}\right)^2}$$

where $D = \dfrac{Et^3}{12(1-v^2)}$. \hfill (3.1)

The stress in the z direction is assumed to be 0 because the material is very small in the z direction compared to the other dimensions. Therefore, only plane stresses are considered. Note that this step is only to validate the suitability of the simulation environment. With other boundary conditions, the stress in the z

direction will not be 0; hence force-induced charges can be calculated. The stresses in the x and y directions and the shear stress (τ_{xy}) in the x–y plane can be found from the following expression:

$$
\begin{pmatrix} \sigma_{xx} \\ \sigma_{yy} \\ \tau_{xy} \end{pmatrix} = \frac{-Ez}{1-n^2} \begin{bmatrix} 1 & v & 0 \\ v & 1 & 0 \\ 0 & 0 & \frac{1+v}{2} \end{bmatrix} \begin{bmatrix} \dfrac{\partial^2 \omega}{\partial x^2} \\ \dfrac{\partial^2 \omega}{\partial y^2} \\ 2\dfrac{\partial^2 \omega}{\partial x \partial y} \end{bmatrix}. \tag{3.2}
$$

Based on the foregoing equations, the closed-form relationship between stress and displacement is depicted in Fig. 3.2. Numerical results from COMSOL are shown together with a closed-form solution in Fig. 3.2. The good alignment validates the accuracy of COMSOL as a simulator.

In the foregoing paragraphs, the theoretical analysis of a simply supported plate was investigated. Owing to the lack of literature on the fully clamped case, simulation results from COMSOL are utilized to analyse the resolution and sensitivity for displacements of the touch panel and stresses on the piezoelectric film. The results for location 1 and 5 of stack-up 1 are illustrated, as stack-up 1 has the lowest panel thickness and consequently the largest displacement among the four stack-ups. Locations 1 and 5 represent two extreme cases. Uniform forces over small concentric circles of radius 1 mm are applied at locations 1 to 5, with different strength levels (0.1 N to 1 N).

The simulation results are depicted in Fig. 3.3. As described previously, five touch locations were investigated, which can be divided into two scenarios. Locations 4 and 5 are at and near the centre region of the panel, and thus can be explained by fully clamped plate theory. In contrast, locations L_1 to L_3 are close to the edges of the plate; hence they are approximated in a better way by axial compression in the z plane [16].

From Fig. 3.3, it can be observed that with the increment of piezoelectric film thickness, the displacements and stresses at locations L_4 and L_5 drop, which aligns with our expectation. This can be generally explained by bending stiffness (K), which is the function of Young's modules (E) and thickness (t). With the increase of E and t, K boosts. However, at locations L_1 to L_3, the displacements increase, together with the piezoelectric film thickness. This is because when the edges of the plate are fixed, the displacement is due purely to the compressive strain in the z direction. As the thickness goes up the strain will therefore increase because the piezoelectric film has a lower Young's modulus than the glass. This also explains the stress resolution results. As the frames support the whole panel, most of the stress is concentrated at the frame regions.

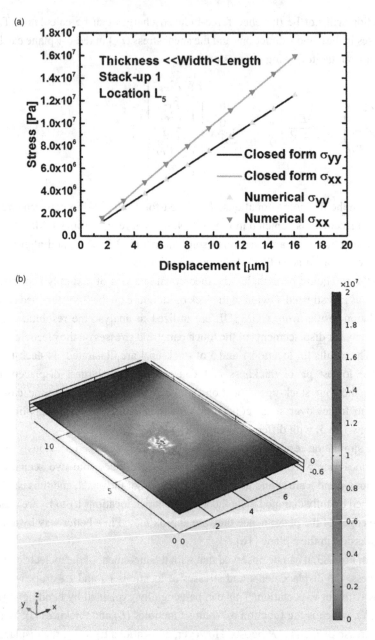

Figure 3.2 (a) Stress versus displacement when thickness is 0.5 mm at location L5. (b) Stress [Pa] distribution of the touch panel. In the simulation, only one layer of PVDF is used; hence the displacement contributed by normal strain is much smaller than the curvature.

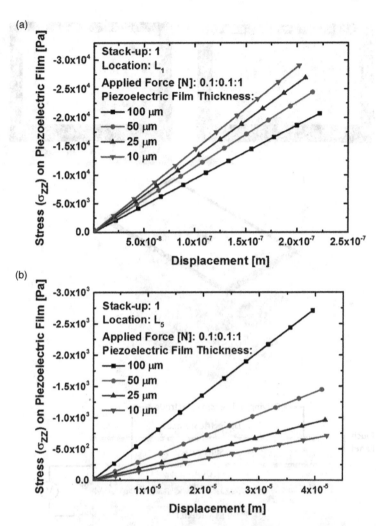

Figure 3.3 Resolution results for stack-up 1: displacement and stress of stack-up 1, location L1 (a) and stack-up 1, location L5 (b). 0.1:0.1:1 indicates from 0.1 N to 1 N, with a step of 0.1 N.

That is the reason why the stress values of location L_4 and L_5 are much smaller than those at locations L_1 to L_3, while location 1 has the highest stress value.

To examine further the proposed technique, a layer of piezoelectric film (d_{33} = 20 pC/N) is laminated with a commercial touch panel (Microchip Inc.). Optical clearance adhesive (OCA) glue is used for the lamination. The original touch panel and laminated touch panel are shown in Fig. 3.4a and b. The structure and working principle of the laminated touch panel are shown in Fig. 3.4c and d. The touch panel's electrodes and the ITO-coated PET consist

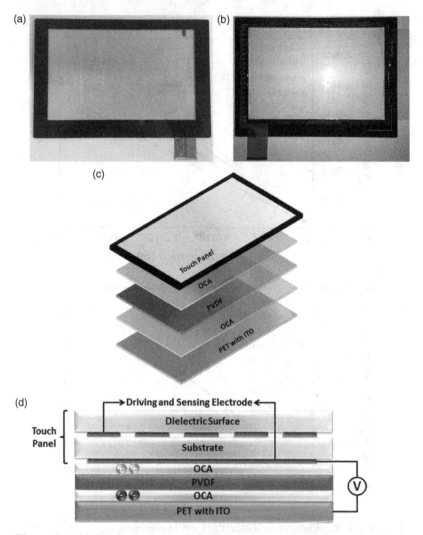

Figure 3.4 (a) Original touch panel. (b) Touch panel laminated with PVDF- and ITO-coated PET. (c) Structure of the laminated touch panel. (d) Working principle of collecting force induced charges.

of many capacitors. By selecting the correct electrode pin, the force-induced voltage can be measured by an oscilloscope. When a 1 N perpendicular touch happens at the centre of the panel edge, the output signal peaks at 0.2 V (as shown in Fig. 3.5), indicating a high force sensitivity. There is a non-uniformity issue in force–voltage responsivity for the proposed structure, indicating different touch locations are associated with different responsivities. However, this can be calibrated by using the capacitive information, which includes the touch location.

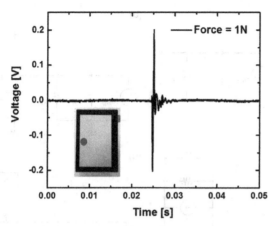

Figure 3.5 Voltage response of a 1 N perpendicular force at the center edge of the piezo film laminated touch panel. The red point indicates the touch location. The actual contact area is 1 mm².

From the simulation and experimental results, a good responsivity can be achieved at 0.42 V/N. However, noise could affect signal detection, especially the system's sensitivity. Therefore, an electrical analysis in terms of SNR has been carried out.

3.2 Electrical Analysis of Proposed Touch Panels

The force applied to the touch panel needs to be converted into an electric signal for the processor to manipulate. Henceforth, a transimpedance amplifier (TIA) is employed as the readout circuit. (A readout circuit should ideally have very low input impedance for collecting charge from the piezoelectric layer. Thus, the transimpedance amplifier is the good solution because its input illustrates a virtual ground to the piezoelectric signal. Furthermore, any force-induced charge response is more linear than the voltage response, from the standpoint of change in distance between electrodes by virtue of strong force touch events.). The SNR is an important parameter for a sensing system. In this section theoretical analysis of the SNR value for the presented piezoelectric film based force sensing system is performed.

Piezoelectric films can be modelled as a charge generator in parallel with a capacitor (C_{PF}) and an internal film resistor (R_{PF}) [17] as depicted in Fig. 3.6a. Because different circuit designs contribute to various noise components, a typical charge amplifier is analyzed as an example. This is shown in Fig. 3.6b.

As the piezoelectric force sensor and readout circuit are uncorrelated noise sources, the input-referred noise power $\overline{i_{IN}^2}$ of the force sensing system is expressed as

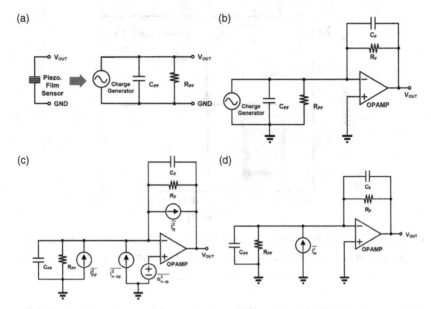

Figure 3.6 (a) Equivalent circuit of piezoelectric film–based force sensor. (b) Charge amplifier based readout circuit. (c) Noise sources of the circuit. (d) Input referred noise source includes all the noise sources.

$$\overline{i_{IN}^2} = \overline{i_{PF}^2} + \overline{i_{RC}^2}, \tag{3.3}$$

where $\overline{i_{PF}^2}$ and $\overline{i_{RC}^2}$ are the noise power spectrum densities (PSDs) of the piezo-electric touch sensor and the readout circuit, respectively. From Fig. 3.6c, $\overline{i_{RC}^2}$ includes feedback resistor noise $\overline{i_R^2}$ and the input current and voltage noise of the opamp ($\overline{i_{n-op}^2}$ and $\overline{e_{n-op}^2}$, values can be retrieved from operational amplifier product datasheet) [92]. $\overline{i_R^2}$ is expressed as

$$\overline{i_R^2} = \frac{4kT}{R_F}, \tag{3.4}$$

where k is the Boltzmann constant and T is the temperature in degrees Kelvin. Note that $\overline{i_{n-op}^2}$ is the shot noise of the input devices. The PSD of the output noise generated by $\overline{e_{n-op}^2}$ is

$$
\begin{aligned}
\overline{v_{on}^2} &= \overline{e_{n-op}^2} \left| 1 + \frac{Z_F}{Z_{PF}} \right|^2 \\
&= \overline{e_{n-op}^2} \frac{(R_{PF} + R_F)^2 + 4\pi^2 f^2 R_F^2 R_{PF}^2 (C_F + C_{PF})^2}{R_{IN}^2 (1 + 4\pi^2 f^2 R_F^2 C_{PF}^2)}.
\end{aligned}
\tag{3.5}
$$

Here Z_F and Z_{PF} are the feedback and piezoelectric sensor impedance values. As R_{PF} is much higher than R_F, Eq. (3.5) is simplified as

$$\overline{v_{on}^2} = \overline{e_{n-op}^2} \frac{1 + 4\pi^2 f^2 R_F^2 (C_F + C_{PF})^2}{(1 + 4\pi^2 f^2 R_F^2 C_{PF}^2)}. \tag{3.6}$$

If a CMOS-based operational amplifier is utilized, the shot noise can be neglected. Hence, the total input-referred noise PSD is expressed as (Fig. 3.6d)

$$\overline{i_{IN}^2} = \frac{4kT}{R_{PF}} + \frac{4kT}{R_F} + \overline{e_{n-op}^2} \left[\frac{1}{R_F^2} + (2\pi f)^2 (C_F + C_{PF})^2 \right]. \tag{3.7}$$

If the force is assumed to be linearly increased within a 'press' period, assumed to be 0.1 s (Δt), and that the signal bandwidth (f_S) is 10 kHz. Then the signal energy (E_{Signal}) is calculated as

$$E_{Signal} = \frac{Q_T^2}{2C_{PF}f_S}, \tag{3.8}$$

and the power of the signal is $E_{Signal}/\Delta t$.

Here Q_T is the total generated charge during the 'press' period. If a lower frequency band is needed, a suitable R_F can be selected. For example, when the frequency goes to 1 Hz, a 100 MΩ resistor could be used, and the lower cut-off frequency is around 0.159 Hz. The calculations yield an SNR of 36.4 dB, which indicates good detection accuracy and a slightly higher value than the measured experimental result of 33.5 dB (by using the same sample for the power spectral density as in Fig. 2.13).

Besides the SNR, another thing that should not escape one's attention is the electrostatics analysis. As the substrate of the touch sensors is changed from glass to piezoelectric film, the mutual capacitance (C_M) between touch panel's driving and sensing electrodes is decreased with the drop of dielectric coefficient. For force sensing, it is necessary to consider if the applied force will change the distance between the electrodes to further affect the capacitance measurement, as the Young's modulus of the piezoelectric film is much smaller than that of glass. In the simulation, a 5 N force is applied at location 5 of the dielectric surface (panel glass). The capacitance (7.76 pC) remains constant before and after the applied force, indicating that the capacitance change is negligible.

3.3 Separation of Force and Capacitive Combined Touch Signals

In previous sections, four widely used stack-ups for capacitive touch panels are proposed for piezoelectric material–based force sensing, and a detailed analysis for both mechanical and electrical aspects is provided. Based on the theoretical analysis and simulation results, we learned that the piezoelectric material–based force sensing technique can provide high force detection sensitivity. However, as the electrodes in the touch panel are shared by the force touch and capacitive touch signals, strong interference from each will decrease the accuracy of both capacitive and force touch detection. Thus interference elimination techniques must be applied to avoid the drop in detection accuracy.

To separate these two external stimuli, analysis of the frequency property of each is expected. The frequency band of the capacitive touch signal is normally around 100 kHz [98]. For example, in nexus 7, the working frequency for the capacitivetouch signal is 90 kHz [62]. No literature is available as to the working frequency for a piezoelectric material–based human force touch signal. Thus, in Section 2, discussions and a simple experiment were carried out to find out the frequency band of the force touch signal. The experiment demonstrated that the force touch signal is a low-frequency signal that occupies the frequency band from 0 to 10 kHz. The working frequency bands are conceptually shown in Fig. 3.7a. Thus, by using a low-pass filter and a band-pass filter, these two signals can be smoothly separated. As shown in Fig. 3.7b, the low-pass filter and the band-pass filter can be implemented on the circuit or digital signal processing (DSP) side by using algorithms. However, it is also noted that the force touch–induced signals are impulses, indicating that the cut-off frequency of the low-pass filter should be much higher than 10 Hz. This will be demonstrated in the text that follows by experimental results.

Intuitively, these two signals (capacitive and force touch signal) should be added at the source, and then separated by the filters as discussed earlier. The combined signal can be assumed as a high-frequency signal modulated by a low-frequency signal. To demonstrate this, a function generator is used to generate a stable high-frequency signal that represents the excitation signal for capacitive signal sensing. A commercial piezoelectric based touch sample (PEDOT/PVDF) [99] is used for outputting the force signal. The fake capacitive signal and the real force signal are sent to the different channels of an oscilloscope (channel 1 and channel 2), and then these two signals are added together using the math function of the oscilloscope. The signals are shown in Fig. 3.8a and b. In Fig. 3.8, the yellow signal is the force signal, which is relatively low frequency, and the green signal represents the capacitive signal, whose frequency is 100 kHz. The blue signal is the combined signal, which is added up by the oscilloscope. The signals in Fig. 3.8a and b are identical, with the only difference between these two figures being the time scale. The result aligns with the previous expectation.

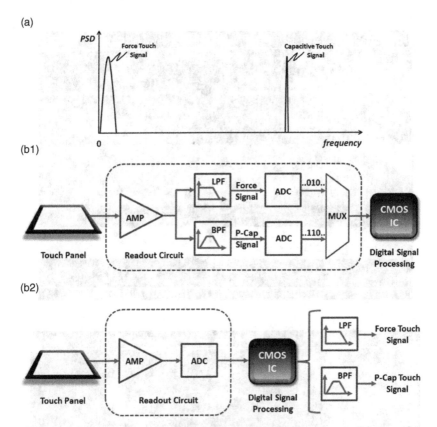

Figure 3.7 (a) Frequency bands of force touch signal and capacitive touch signal. (b) Block diagram of low-pass and band-pass filtering based force touch and capacitive touch signal interpretation circuits. Filtering is implemented by (b1) hardware and (b2) software).

The next step is to implement the foregoing signal combination in a circuit. In Section 2, different methods for capacitive signal sensing were introduced. An easily implemented method is needed that can detect the force touch signal and capacitive touch signal at the same time. To satisfy the two requirements, it was decided to use a charge amplifier. The circuit for combining capacitive and force signals is shown in Fig. 3.9. From the circuit diagram it can be seen that the value of the capacitance C_{Sensor} is related to the piezoelectric material–based touch sensor. However, when a human finger taps on the touch panel, the value of the C_{Sensor} increases, because the finger-related capacitance is in parallel to the original touch sensor, boosting the capacitance value [98]. The relationship between charge, voltage, and capacitance,

$$Q = CV, \tag{3.9}$$

(a)

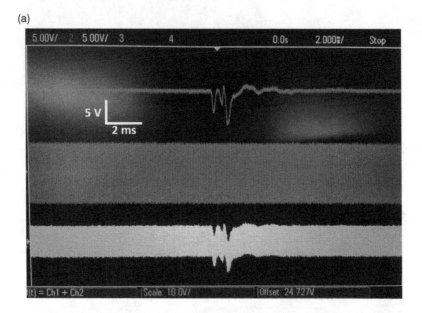

(b)

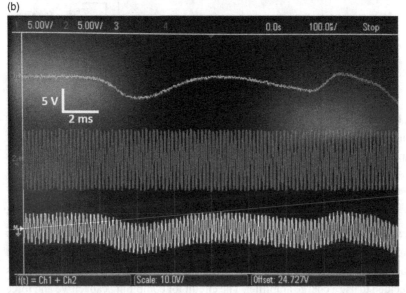

Figure 3.8 Capacitive touch signal and force touch signal and their combination in a time domain.

shows that when the capacitance value increases for a given voltage level, the amount of stored charge decreases with linearly. Thus by periodically measuring the stored charges, we can know the capacitive change and hence interpret the presence of the capacitive touch events.

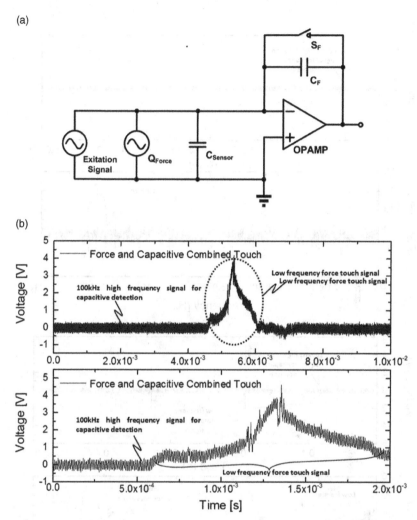

Figure 3.9 (a) Readout circuit for force and capacitive touch signals. (b) Capacitive and force touch combined signal in time domain.

The same experiment was performed with the charge amplifier. The output was illustrated using the oscilloscope, and showed that the force signal modulates the capacitive signal, as expected. Figure 3.9a and b shows the same signal with different time scales, to clearly demonstrate this phenomenon. To further investigate how to separate these two signals, the force and capacitive combined signal is shown in the time domain and frequency domain. Figure 3.10a shows that when no touch events occur, there is a signal in the frequency domain only at 100 kHz. In contrast, when touch events are performed, energy is boosted within the 0 to 10 kHz range, indicating that the

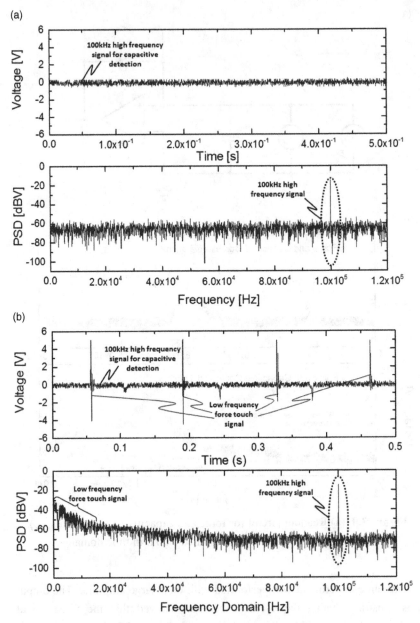

Figure 3.10 Frequency bands of force touch signal and capacitive touch signal.

force signal occupies the low-frequency range. As the frequency bands of force and capacitive touch signals are close to each other (only one order), applying filters is not easy using hardware; thus it is suggested that digital filters be applied.

3.4 Conclusion

In Section 2, a sandwiched multi-functional stack-up was proposed after reviewing capacitive touch panels and techniques related to piezoelectric materials. In this section, the mechanical and electrical characteristics of the multi-functional based capacitive touch panel are investigated. Based on the analysis, the proposed technique can provide better force touch accuracy and sensitivity. Furthermore, the force touch signal and capacitive touch signal can be separated in the frequency domain, validating the possibility of concurrently detecting multi-dimensional signals. The work presented in this chapter provides design considerations for implementing piezoelectric materials into commercial capacitive touch panels.

4 Fabrication and Measurement of Flexible Multi-Functional Touch Panel

In Section 3, the proposed multi-functional touch panel was evaluated theoretically. The analysis shows that the proposed technique can provide higher force detection accuracy and sensitivity compared to current commercial 3-D (x–y–z) touch panels [28], [29]. The analysis was further proven by laminating a PVDF/ITO/PET sandwiched layer with a commercial capacitive touch panel. In this section, the proposed flexible multi-functional touch panel in Section 2 will be fabricated and measured.

Here, mono-layered graphene [100]–[110] is used as a flexible and transparent electrode, due to its high optical transparency (97.7%), low sheet resistance (30 Ω/Υ for highly doped and 300 Ω/Υ for undoped [110]), and high mechanical strength (Young's modulus of 1 TPa and intrinsic strength of 130 GPa [109]). In particular, the fracture strain of graphene can be more than 10 times higher compared to ITO [110] (ITO: 0.003–0.022 [111], graphene: 0.14 [109]), indicating that graphene is a strong candidate for use as an electrode. In addition to the foregoing reasons, we also want to prove that our proposed technique can cover a variety of materials, from conventional (e.g. ITO) to advanced (e.g. graphene) materials.

4.1 Touch Panel Fabrication

Graphene is obtained by the previously reported chemical vapour deposition (CVD) recipe [100], the process for which is shown in Fig. 4.1. First the graphene is grown on the 5 × 5 cm^2 copper substrate, which is then etched using acid (graphene is patterned by using oxygen plasma). After the CVD process, the synthesized graphene needs to be transferred from the Cu substrate onto the PET substrate. In this work the conventional wet transfer method is

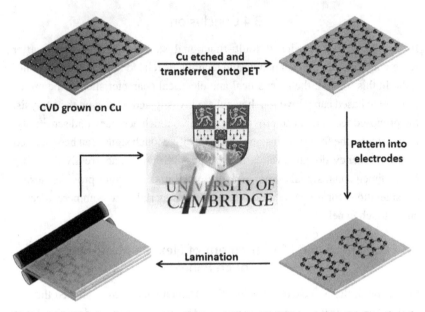

Figure 4.1 Fabrication process of the graphene/PVDF based multi-functional touch panel.

employed [100], [112], [113]. First the graphene/Cu layers are spin coated with a thin layer of the poly(methyl methacrylate) (PMMA) A4 950 photo-resistance (~300 nm thickness) at a rotation rate of ~4000 rpm for ~ 40 s, for the purpose of protecting the graphene in subsequent processing. The sample is treated by a mild oxygen plasma in a reaction ion etching (RIE) system. The graphene grown on the backside of the Cu foil without the protection by PMMA is removed by the plasma. Next the PMMA/graphene/Cu stack is floated on 0.05 M aqueous ammonia persulfate (APS, (NH4)2S2O8), which gradually etches the Cu, and then a rinse step is repeated three times. Next the sample is hooped onto the target substrate, which is then left overnight in air to gently dry the residual water. This way the graphene film is tightly bound to the substrate. Finally the dried sample is immersed in acetone overnight to remove the PMMA coating. The above procedures allow the graphene film to stick to the PET substrate. The foregoing procedure is repeated to produce two graphene/PET structured devices. One of the devices is used as the bottom (ground) layer, and another device is patterned, working as the top electrode layer. The two devices are laminated with a piece of commercial β-phase PVDF from Solvay Corp., becoming a flexible, transparent and multifunctional touch panel (structure is shown in Fig. 4.2). Lamination is processed in 90-degree environment; the pressure is not measured because of a lack of technique support. The high temperature degrades the polling process of the PVDF film, which could

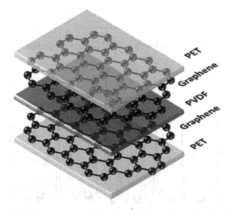

Figure 4.2 Structure of the multi-functional touch panel.

weaken the β-phase structure of the film and degrade the piezoelectric response. Hence, the obtained value of force–voltage responsivity would be smaller than expected.

The fabricated touch panel is expected to be sensitive to changes in capacitance and force. To measure the change of capacitance, a parameter analyser (Keithley 4200 SCS) is used. Two scenarios of capacitive touch are performed and measured. In the first scenario conventional tapping touches are performed, indicating that the finger contacts the surface of the touch panel. In the second scenario, different layers of microscope slides (Thermo Fisher Scientific Inc.) are used to mimic a variety of distances for hover touch events. The thickness of each microscope slide is around 0.5 mm.

4.2 Touch Panel Measurement

4.2.1 Capacitive Response to Human Finger Touch

As mentioned earlier, two types of capacitive touch events are performed: the tapping (contact) touch and hover (non-contact) touch. For the tapping touch, the amplitude of minimum capacitive change is measured. This is because a touch panel system normally has a threshold set to determine whether or not a contact touch event happens. The minimum capacitance change can help in determining the threshold of registering the touch events. It should be noted here that the minimum capacitance change is not the minimum detectable capacitance change, which is related to many factors such as the AC measuring signal frequency and the noise floor.

A series of contact touches were performed; part of the experimental results is shown in Fig. 4.3a. As explained earlier, the intention of this experiment was to help estimate the touch detection threshold. Thus, the strength of the performed

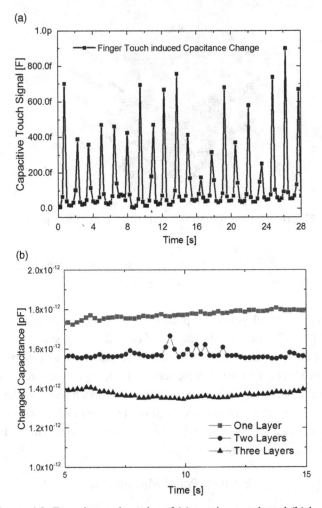

Figure 4.3 Experimental results of (a) tapping touch and (b) hover touch–related events.

contact touches must be similar to that of conventional tapping touches with touch panels. The experimental results show that the minimum capacitance change caused by the finger contact touch was around 200 fF, which can be considered to be a threshold. Two factors related to the different capacitance values induced by the contact finger touches are the contact area (A) and the distance. The contact area has a positive relationship with the applied force. In contrast, the distance between the finger to the sensing electrode (d_1) and the distance between the sensing electrode and the driving electrode (d_2) have a negative relationship with the strength of the applied force.

The precise reduced distance value depends on the applied force, mechanical properties of the touch panel (such as Young's modulus and Poisson ratio of each layer), and boundary conditions (such as simply supported and fully clamped).

In terms of non-contact touches, different layers of plain microscope slides were used to control the distance between the finger and the touch panel. Each glass slide is around 0.5 mm thick. The experimental results are shown in Fig. 4.3b. It can be observed that the hover touch–induced capacitance change decreases as the number of glass slides increases. The number of glass slides and the change in capacitance value are negatively correlated, which aligns with our expectation. The detailed experimental results of hover touch events are illustrated in Fig. 4.4. The experimental results demonstrated in Figs. 4.3 and 4.4 show that the fabricated touch panel can detect both contact and non-contact touch events, satisfying the needs of customers. The fluctuations in Figs. 4.3b and 4.4 are caused by the unstable finger position. The differences in measured results from the four touch pads are within 5%.

4.2.2 Force Response to Machine Stylus Touch

To examine the performance of the fabricated touch panel in terms of its response to the force touch, a test-bed was built as shown in Fig. 4.5a. It consists of three main components: a 3-D position control system for the touch panel (Fig. 4.5b), a stable force source (shaker), and an accurate force sensor (500 mV/lb) (Fig. 4.5c).

The 3-D position control system consists of three motors used to accurately control the position of the touch panel along the $x–y–z$ axis. The resolution of the 3-D positioning system is 1 mm. A LabVIEW based graphic user interface (GUI) (Fig. 4.5d) was designed to control the positioning system. After the touch panel was positioned properly, the shaker was started to provide a stable force touch signal at 10 Hz. The stick of the shaker can be tuned from 0 to 90 degrees. In our measurement, the direction of the stick was fixed at 90 degrees (perpendicular) to the fabricated touch panel. The strength of the force touch signal can be read by the force sensor mounted on the stick of the shaker, as shown in Fig. 4.5c. By comparing output signals from the force sensor and the touch panel, the force/voltage coefficient of the fabricated touch panel could be calculated.

There are some limitations to the force touch measurement system. First, because the material of the plate holding the touch panel is conductive, strong

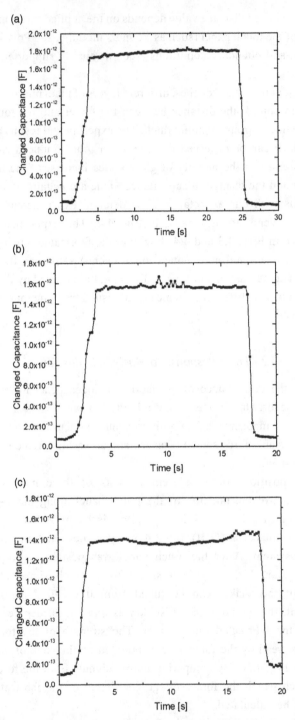

Figure 4.4 Experimental results of hover touch events with (a) one, (b) two, and (c) three layers of plain microscope slides.

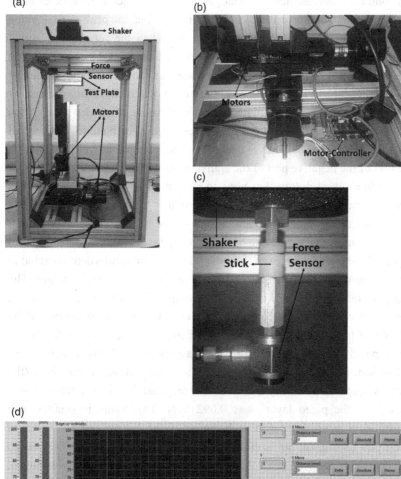

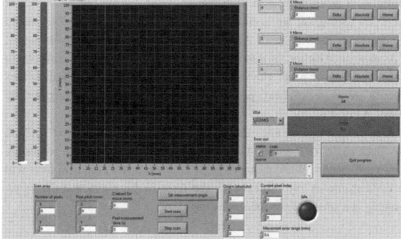

Figure 4.5 (a) Test-bed for force touch experiment, and (b) three-motor controlled positioning system. (c) Shaker stick mounted with a force sensor. (d) LabVIEW based GUI for the 3-D positioning system.

environmental interference (e.g. charger noise) is coupled and hence disturbs the force touch measurement. As shown in Fig. 4.6a, the peak to peak value of the charger noise can even exceed 1 V. When the force touch signal combined with the charger noise, it was difficult to read the accurate value (shown in Fig. 4.6b). To solve this, the plate was connected to the ground of the oscilloscope. Fig. 4.6c shows the result when the charger noise is cancelled out. Second, the plate is not fully fixed. When the stick of the shaker hits the touch panel, the plate moves together with the stick in a small distance. When the stick moves back, the plate rebounds as well, resulting in the loss of the negative part of the signal, as shown in Fig. 4.6c. This will be solved in a future design. For now, to keep the stage stable, the plate was manually held, and the frequency of the shaker was controlled to be 1 Hz (note, due to the functional limitations of the shaker, the shaker cannot provide a sinusoidal signal at a low-frequency range. Instead of a sinusoidal signal, the shaker produces an 'impulse' signal). The results demonstrated in Fig. 4.7 show that both positive and negative components were obtained. The signal from channel 2 in Fig. 4.7 is the output from the commercial force sensor. By comparing the outputs from the commercial force sensor and the fabricated touch panel, the piezoelectric d_{33} coefficient of the fabricated touch panel can be obtained. In our measurement, 100 experiments were performed, and the averaged piezoelectric force–voltage responsivity (the force–voltage responsivity indicates the responsivity for the whole touch panel, not the piezo layer) was 0.092 V/N. This value is smaller than expected; a possible reason is because of the piezoelectric coefficient decreases during the lamination process. In lamination, high temperature (90 degree) is applied; hence the piezoelectric property is slightly damaged.

As mentioned earlier, the direction of the stick was set to be perpendicular to the fabricated touch panel. However, in practical touch scenarios, the directions of touch events are not always perpendicular to the touch panel, as conceptually shown in Fig. 4.8.

To investigate how the angle of a touch event affects the touch panel's output, two touch angles (80° and 70°) were performed. The experimental results are shown in Table 4.1. When the touch angle is not normal, a portion of force will be divided into an x–y plane; hence the force in the z direction decreases, resulting in a smaller force-induced voltage.

4.3 Conclusion

In this section, the proposed multi-functional touch panel was fabricated and measured. CVD grown mono-layer graphene was used as the electrodes, which

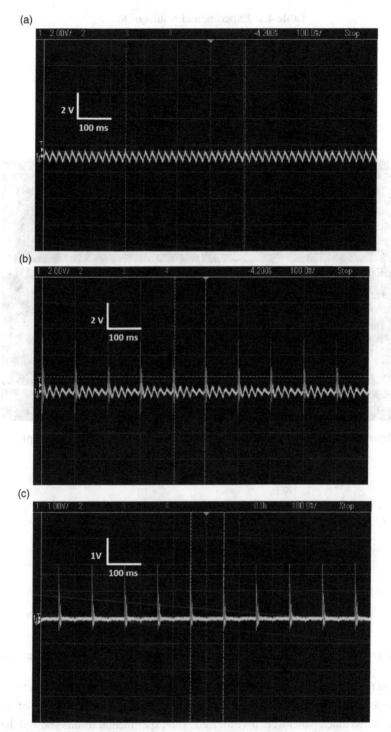

Figure 4.6 Experimental results of (a) charger noise, (b) force touch signal with charger noise, and (c) force touch signal after charger noise cancellation.

Table 4.1 Experimental results of touch
events from different directions

Touch angle	V/N
90°	0.092
80°	0.084
70°	0.069

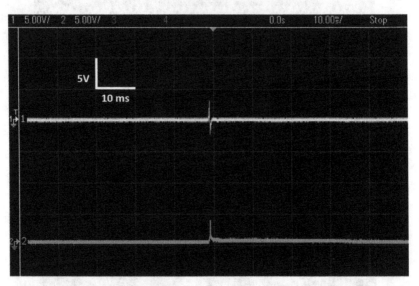

Figure 4.7 Experimental results of force touch signal with 'tabilized' plate.

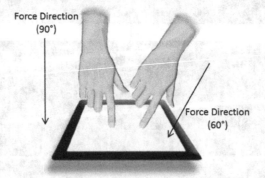

Figure 4.8 Conceptual descriptions of touch events from different directions.

were then laminated with a commercial PVDF film to form the multi-functional
touch panel. The capacitive touch response and force touch response of the
fabricated touch panel were investigated. The experimental results showed that
the fabricated touch panel can successfully detect capacitive and force touch

events with high sensitivity (0.2 pF and 92 mV/N) and good flexibility, demonstrating its potential use for novel applications.

5 Algorithms for Force Touch Signal

A theoretical analysis of the mechanical and electrical properties of the proposed technique was performed in Section 3 in order to validate it, following which a graphene-based touch panel was fabricated and tested in Section 4. From the experimental results illustrated in Section 4, it was learned that the fabricated touch panel can provide good force and capacitive touch detection sensitivity. Section 3 demonstrated that the capacitive touch signal and force touch signal can be separated by using their frequency properties. More specifically, the force touch signal takes the low-frequency range (within 10 kHz), while the capacitive signal occupies a relatively high frequency range of 100 kHz. After separating these two signals, the next step is to interpret them. As detection of capacitive touch signal has been widely studied, this section focuses on developing algorithms to interpret the force touch signal.

First, the strength of the applied force touch signal is addressed. Here the force touch signal can be divided into two scenarios: dynamic force touch events and static force touch events. As a piezoelectric material, PVDF can detect dynamic force touch signals [72], [73], [114]–[123]. However, it is unable to detect a static force signal, as explained in Section 2. This section designs interpretation algorithms to realize both dynamic and static force touch sensing.

Second, stress propagation is a practical issue for the PVDF-based touch panel in detecting force touch events. More specifically, the stress caused by a force touch in one location can propagate to adjacent locations. Although the propagated stress can be small, it is difficult for the system to distinguish whether the detected signal at a specific location is generated from a real touch event at this location, or propagated from other force touch locations. The algorithm developed in this section overcomes this issue as well.

Third, power consumption is a critical issue in electronics. In the later part of this section, an algorithm is proposed and demonstrated that can concurrently detect force touch signals and harvest force touch generated electrical energy.

5.1 Algorithms for Force Touch Interpretation

As mentioned in Section 4, one piece of graphene electrodes was patterned into four small square areas, indicating four touch sensors. The other piece of graphene was maintained and works as a ground layer. PVDF film was settled in between the two graphene layers, functioning as an insulating layer for the

(a)   (b)

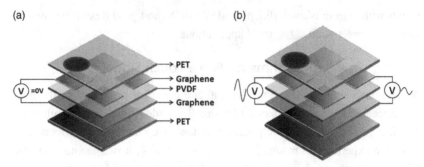

Figure 5.1 Two issues of piezoelectric material–based force touch sensing in a touch panel. (a) Static force touch is not detected. (b) Propagated stress–induced charge may result in force touch misregistration. The red circles indicate the force touch positions. The forces are static force and dynamic force in (a) and (b), respectively.

graphene based capacitor and the force detection and energy harvesting layer by generating force-induced charges. One obvious drawback of employing piezo-electric phenomenon of PVDF, or any other non-centrosymmetric material, for force touch detection is its inability to recognize static force due to thermal drift and current dissipation [117]. In addition to that, when a force touch is performed at one location, mechanical stress can be transferred to adjacent areas. The amount of propagated stress depends on the mechanical properties (e. g. Young's modulus) and boundary condition (e.g. simple supported) of the touch panel, and the character of the applied force touch. Although the trans-ferred mechanical stress and induced charges are sometimes small, it is difficult to distinguish whether they have been induced by a light touch or by an adjacent heavy touch. Hence, the propagated stress may give rise to force touch mis-registrations. These two issues are shown in Fig. 5.1.

To solve these two issues, capacitive signals are used. For the former, the existence of a capacitive signal can be used to indicate a static force signal, as these two signals are generated together by a force finger touch. Thus, although the force signal may fall to zero, or below a certain threshold, as long as the capacitive touch signal is detected, the force touch is assumed to be maintained at the same level as its peak value. This is conceptually shown in Fig. 5.2. In Fig. 5.2a, the static force touch lasts within the time period t_1 to $t_1 + \Delta t_2$; however, the force touch signal is detected only during the time period t_1 to $t_1 + \Delta t_1$. After applying the algorithm, if a capacitive touch signal is detected, then the force is assumed to be maintained.

As to force touch misregistration due to propagated stress from adjacent force touches, the capacitive touch signal is also employed to distinguish a real finger

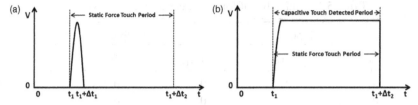

Figure 5.2 Conceptual output of the static force detection algorithm.

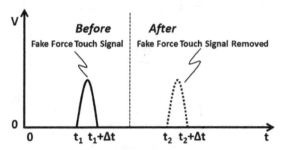

Figure 5.3 Conceptual outputs of the fake force touch signal before and after the algorithm.

force touch signal from a fake one. This is achieved based on the fact that when a location experiences propagated stress from adjacent force touches, its capacitance does not change dramatically. Thus, if there is no capacitive touch signal at a specific location, we assume that there is no finger-based force touch, even if the readout shows an observable force signal. This is conceptually described in Fig. 5.3. A truth table of force and capacitive touch signals is provided in Table 5.1, and the algorithm flow chart is described in Fig. 5.4.

To implement the designed algorithms, a touch panel system was assembled as shown in Fig. 5.5a and b. First the static force touch interpretation algorithm was applied. The force touch and capacitive touch signal output from a single channel is shown in Fig. 5.6a and b. Figures 5.6a and b show that static force touch is detected, and that the force is maintained during the static force touch period. A larger capacitance signal is obtained when a stronger force touch occurs. This is because when a stronger finger touch occurs, normally the contact area between the touch panel and human finger becomes larger, resulting in a bigger capacitive signal. In Fig. 5.6c and d, outputs from two adjacent channels are plotted. The figure shows that, when a force touch occurs at sensor 1, the propagated stress can give rise to a light force signal at sensor 2, resulting in potential force touch misregistration. In contrast, the capacitive signals at the adjacent sensors are very small, and thus can be used as determination signals

Table 5.1 Truth table of force and capacitive touch signals

Input		Output	
$T_{Capacitive}$	T_{Force}	$T_{Capacitive}$	T_{Force}
0	0	0	0
0	1	0	0
1	0	1	0
1	1	1	1

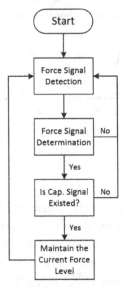

Figure 5.4 Flow chart of the force detection algorithm.

for force touch detection. After application of the propagated stress elimination algorithm, the issue is solved as shown in Fig. 5.6e.

5.2 Energy Harvesting

Energy consumption is a practical and critical issue in today's electronic devices. In smart phones, high power consumption, caused mainly by 3G (~31%), display (~24%), and Wi-Fi (~24%) [123] components, shortens the lifetime of the battery, giving rise to the popularity of portable charging devices [47]. Many low-power consumption techniques have been proposed and implemented [63], [124]–[128]. In addition to these, energy harvesting techniques are attracting attention. Traditional energy-harvesting techniques in smart phones are based on collecting RF, solar, and thermoelectric related energy. However,

(a)

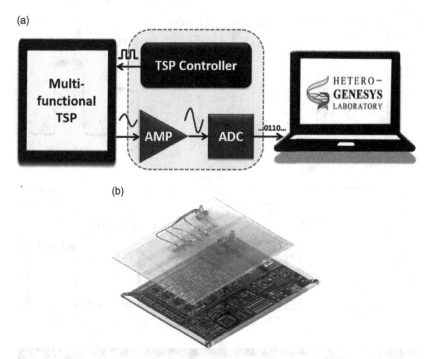

(b)

Figure 5.5 (a, b) System diagram and interface circuit.

harvested RF energy is only approximately $0.1~\mu W/cm^3$ [129], strong sunlight is needed for collecting solar energy, and thermoelectric energy requires stable heat sources, which are not convenient for customers [129]. Thus, a green and convenient energy harvesting technique is required for smart phones.

As mentioned previously in Section 2, the force-induced charge can be employed to interpret the force level, or be stored for future use. The previous section demonstrated how the force induced charge is used for distinguishing the force level. This section provides a simple energy harvesting system, and the harvested energy is then used to light a blue LED.

5.2.1 Piezoelectric Material–Based Energy Harvesting System

As conceptually shown in Fig. 5.7, the piezoelectric material–based force-induced electric signal has positive and negative components because the polarization changes twice during a whole press and release finger touch procedure. This kind of signal cannot be directly stored in an energy storage component such as a capacitor, because the negative part will compensate for the positive part. A single-diode–based rectifier circuit can be used to block the negative part, as shown in Fig. 5.8a. However, blocking the negative part wastes half of the mechanical (force) energy transformed to electrical energy. Thus, a

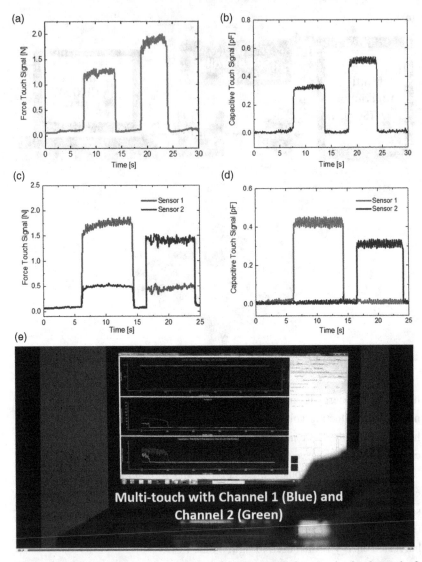

Figure 5.6 (a, b) Force and capacitive signals output from a single channel of the system. (c, d) Force and capacitive signals output from two adjacent channels of the system before the propagated stress elimination algorithm is applied. (e) Panel capture of the software.

bridge rectifier circuit as shown in Fig. 5.8b is employed to fully collect the energy. The conceptual output after the bridge rectifier is described in Fig. 5.8, and the collected energy in the capacitor is shown in Fig. 5.9. In Fig. 5.9a, the charging and discharging periods are clearly shown. After each force touch the capacitor is charged by a little bit. Zooming-in on the charging period shows

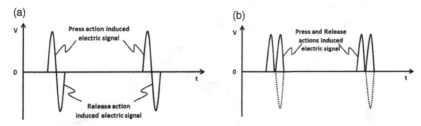

Figure 5.7 Conceptual piezoelectric based finger touch signal.

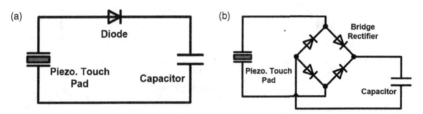

Figure 5.8 (a) Single-diode–based rectifier circuit. (b) Four-diode–based bridge rectifier circuit.

that there are two charging moments during a single force touch. Because the polarization alters twice, the energy in the negative part is also collected by the bridge rectifier, and used to charge the capacitor.

The foregoing demonstrates that force-induced charges can be used to charge a capacitor; thus, when the software application doesn't need to read force information, the force-induced charge can be used for energy harvesting. It is also possible to read the force signal and do energy harvesting at the same time. For example, the positive part can be used for force signal detection, and the negative part used for energy harvesting, as shown in Fig. 5.10.

Figure 5.10a shows that the force detection circuit and energy harvesting circuit share the same ground. Both use half of the energy generated by force touches. To illustrate these two functions, the force detection circuit and the energy harvesting circuit were connected to an oscilloscope, as shown in Fig. 5.10b. The corresponding touch signal and energy harvesting signal are shown in Fig. 5.11, respectively. Force detection and energy harvesting were achieved at the same time. In practical circuit design, a charge amplifier is widely used to read the force touch signal, due to the high impedance of the PVDF sample. As the oscilloscope is already equipped with high impedance, no charge amplifier was used in this demonstration.

To calculate how much total energy is stored in the capacitor, the following equation can be used:

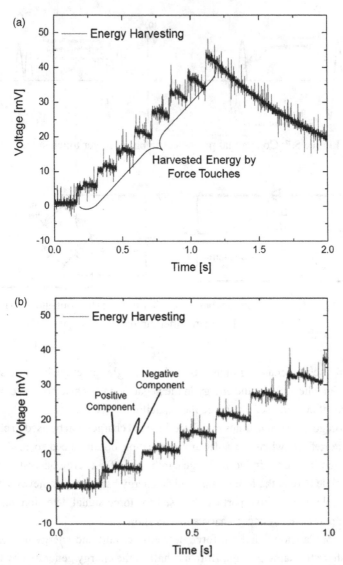

Figure 5.9 (a) Charging and discharging period. (b) Details of the charging period.

$$E_{E.H.} = \frac{1}{2}CV^2, \tag{5.1}$$

where $E_{E.H.}$ is the energy stored in the capacitor, C denotes the capacitance of the capacitor, and V represents the voltage across the capacitor. To calculate the stored energy in Fig. 5.9a, the following values for the variables can be used in Eq. (5.1): C is equal to 1 µF and V is around 45 mV. Thus, $E_{E.H.}$ is 1.03 nJ based

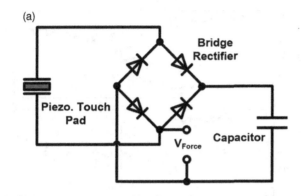

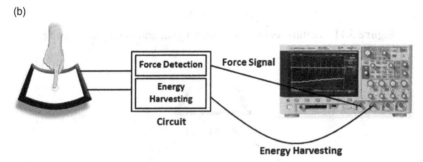

Figure 5.10 (a) Circuit for force detection and energy harvesting. (b) Configuration of the touch panel, circuit, and oscilloscope.

on the equation. To calculate the energy harvested by a single force touch, the following equation can be employed:

$$\Delta E_{E.H.} = \frac{1}{2}C(V' - V)^2, \tag{5.2}$$

where $\Delta E_{E.H.}$ indicates the energy stored by a single force touch, and ΔV represents the voltage change after a single force touch.

Because of the high impedance of the PVDF-based touch panel, the energy generated by the force touch cannot be efficiently transferred and stored in the capacitor. To solve this issue, a Maximum Power Point Tracking (MPPT) circuit is normally used. However, this Element is concerned mainly with demonstrating this function in a touch panel; thus MPPT circuits are not employed.

Based on the work presented in this section, the functionalities in our touch panel expand from two (capacitive and force sensing) to three, including energy harvesting, as shown in Fig. 5.12.

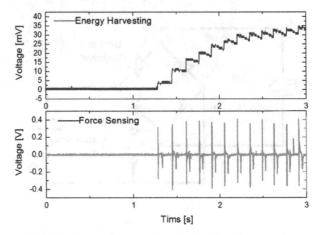

Figure 5.11 Simultaneous force touch signal and energy harvesting.

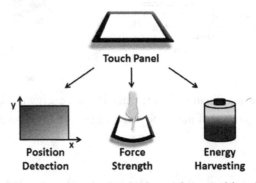

Figure 5.12 Three functions in the touch panel are position detection (by capacitive sensing), force detection, and energy harvesting.

5.3 Conclusion

This section has designed and implemented the interpretation algorithm and energy harvesting algorithms. The force touch interpretation algorithm addresses two issues facing the piezoelectric materials–based force touch panel: static force sensing and stress propagation. The energy harvesting algorithm achieved both energy harvesting and force touch detection by using the altered polarizations of piezoelectric materials during force touch events. Furthermore, the accuracy and resolution for force touch detection are maintained. The content in this section has significant meaning in successfully utilizing piezoelectric material in touch panels for high force touch detection accuracy and enhanced battery lifetime.

6 Conclusion and Future Work

This section summarizes the work provided in this Element, which includes properties of piezoelectric materials and capacitive touch panel–related techniques; a theoretical analysis of the piezoelectric material–based capacitive touch panel, from aspects of the mechanical and electrical properties; and fabrication and measurement of the graphene and PVDF-based multi-functional touch panel and interpretation algorithms for force touch detection. Then, the issues limiting the successful use of piezoelectric material in capacitive touch panels are explained. Finally, this section ends with authors' outlook on the development of the multifunctional touch interfaces.

6.1 Conclusion

This Element has presented a multi-functional capacitive touch panel for multi-dimensional sensing with flexible form factor. The properties of piezoelectric materials, techniques related to capacitive touch panels, theoretical analysis of the mechanical and electrical properties of the multi-functional touch panel, touch panel fabrication and measurement, and interpretation algorithms, have been provided.

Design considerations for detecting force touch in capacitive touch panels were discussed based on the literature review of piezoelectric materials and techniques related to capacitive touch panels. The discussions revealed that the characteristics of capacitive touch panels, such as pre-stress, temperature, and environmental EMI, do not strongly affect the accuracy of force touch detection. In contrast, the characteristics of force touch events, such as force touch speed, touch direction, and finger geometry, can heavily influence the detection accuracy. These design considerations are of significant meaning in designing and calibrating a piezoelectric material–based touch panel.

A theoretical analysis was conducted in terms of the mechanical and electrical properties of the proposed multi-functional touch panel. In the mechanical analysis, four widely used capacitive touch panel stack-ups were investigated. From the investigation, the piezoelectric material–based technique is proved to be able to provide enhanced z-axis touch detection resolution and sensitivity compared to the existing force touch technique, which relies on changes to force-induced capacitance. In the electrical analysis, both the readout SNR and the frequency property of the force touch signal were studied. A high SNR of 59.1 dB was obtained based on the theoretical SNR estimation, indicating high detection accuracy. From analysis of the frequency of the force touch signal, it is demonstrated that conventional tapping force touches occupy the frequency band within 10 kHz, and thus can be separated from capacitive excitation signals, which are normally above 90 kHz in capacitive touch panels.

After the theoretical analysis, a multi-functional touch panel was fabricated. To ensure that the fabricated multi-functional touch panel has good flexibility, graphene electrodes were selected and grown by the conventional CVD method. Two types of measurements (capacitive and force touch measurement) were performed. In the capacitive touch measurement, the minimum changed capacitance value was 0.2 pF, indicating a good detection resolution. In the force touch measurement, the force–voltage sensitivity was 92 mV/N, satisfying applications where high force touch sensitivity is needed.

A touch panel system was then assembled. Two force touch detection issues related to piezoelectric materials were addressed by employing information provided by capacitive touch signals. More specifically, static force touch detection and propagated stress elimination were achieved by incorporating the presence of capacitive touch signals.

Through performing detailed theoretical analysis and assembling a touch panel prototype and illustrating corresponding experimental results, the feasibility of the authors' viewpoint on the developmental trend of the next generation of touch interfaces is validated.

6.2 Challenges and Future Work

Although the prototype demonstrates the use of piezoelectric materials in touch interfaces for multi-dimensional sensing with good flexibility, some challenges remain, preventing the successful use in commercial HMIs for multi-dimensional sensing. This section first explains the challenges and then discusses the potential solutions.

A. Multiple Stimuli Superimposing

A user touch–induced electric signal is a combination of multiple physical phenomena. For example, when a user hits the piezoelectric touch panel's surface, inevitably triboelectrification also occurs. Thus, the detected signal is a combination of piezoelectric and triboelectric effects. However, discrimination of the piezoelectric signal from the combined signal has not yet been reported in the literature. Another example can be given as the combination of piezoelectric and pyroelectric effects, which occurs when ferroelectric materials are used, e.g. PVDF. Although it has been reported that heat transfer is slower than the piezoelectric response, the successful separation of these two effects has not yet been presented in applications for interactive displays. Hence, the precise detection of a single physical effect is demanded to provide customers with higher detection accuracy and to enable potential applications. To address this, we are conducting a deep probe of working principles of these physical phenomena and developing algorithms to eliminate unwanted signals.

B. Unstable Responsivity Induced by User Touch Behaviours

The second challenge stems from the manner of human–machine interactions. Touch events can be carried out by users in a variety of means. For example, a finger touch with the same force amplitude can hit the touch panel from different angles and speeds, with diverse contact areas, resulting in unstable force–voltage responsivity in piezoelectric material–based touch panels, and decreasing the detection accuracy. We are currently solving this by employing artificial neuron networks (ANNs).

C. Lack of Efficient Calibration

A calibration mechanism is vital for most, if not all, sensing systems. However, current studies focus on achieving a high sensing performance (e.g. sensitivity and responsivity) of functional materials–based touch panels, but fail to design calibration functions for maintaining high performance over long-term use. For example, when the mechanical properties of the touch panel change, e.g. due to being occasionally dropped on the floor, the boundary conditions of the touch panel will be altered; therefore the force–voltage responsivity would be changed accordingly for piezoelectric and triboelectric–based touch panels. Although external commercialized force sensors can be used periodically for calibration, it is inconvenient to customers, and we cannot expect customers to purchase additional equipment to use our products. Hence, user-friendly calibration techniques are expected to be developed. To satisfy this goal, we are designing self-calibration techniques that require no external calibration by users.

Abbreviations

AC	Alternating current
ANN	Artificial neuron network
CDS	Correlated double sampling
CS	Compressive sensing
CVD	Chemical vapor deposition
DC	Direct current
DSP	Digital signal processing
EMI	Electro-magnetic interference
HMI	Human–machine interactivity/human–machine interface
ITO	Indium tin oxide
LCD	Liquid crystal display
LPF	Low-pass filtering
MCU	Microcontroller unit
MPPT	Maximum power point tracking
OLED	Organic light-emitting diodes
Op-amp	Operational amplifier
OSA	Optical clearance adhesive
PD	Photon detector
PEDOT	Poly(3,4-ethylenedioxythiophene)
PET	Polyethylene terephthalate
PVDF	Polyvinylidene fluoride
SNR	Amorphous silicon
TFT	Thin film transistor
TP	Touch panel
TIA	Transimpedance amplifier

Appendix

Berlincourt Method for Piezoelectric d_{33} Coefficient Measurement

The Berlincourt method uses the direct effect for quasi-static measurement of the piezoelectric d_{33} coefficient [74]. It is named after Don Berlincourt, who made an important contribution to designing and fabricating the first commercial d_{33} measurement system [72]. A Berlincourt method–based piezoelectric coefficient measurement system normally comprises two units: the force head and the control electronics. The force head consists of the loading actuator and a reference sample that is used for calibration, and the control electronics include the force control (strength and frequency) system, the charge measurement system, and the piezoelectric coefficient (d_{33}) calculation system. There are many factors affecting the accuracy of d_{33} measurement using the Berlincourt method, which are summarized and explained in the text that follows.

AC Measuring Force

The AC measuring force can be described according to its strength (magnitude) and frequency. The magnitude does not significantly influence the measurement result, unless the piezoelectric material works in a non-linear regime. The merit of applying a stronger AC force is to generate more charges, giving rise to a higher signal-to-noise ratio (SNR). However, this may result in the piezoelectric material operating in a non-linear regime.

In contrast, the frequency does result in different d_{33} measurement results. Because of thermal drift and charge dissipation, a static force cannot be applied. The frequency range is approximately 10 Hz to 1 kHz, governed by the charge measurement system, the stability of the generated charge during the measurement, and the load application method. Furthermore, the measurement frequency should avoid the main power frequencies and corresponding harmonics, because it is normally difficult to completely shield electro-magnetic interference (EMI) from surrounding electronic instruments.

Besides the system performance, the effect of the AC force frequency depends on the properties of the tested piezoelectric material. In [75], the measurement results for hard (PC4D) and soft (PC5H) materials were different. The piezoelectric d_{33} coefficient of the hard material was boosted with the increment of AC force's frequency, while the d_{33} of the soft material was stable under 150 Hz, and then followed the same trend as the hard material. This can be explained by the suppression of domain movement at increasing frequencies for soft materials, and the de-aging effect for hard materials.

Effect of Static Pre-load

Static pre-load is used to clamp the samples at a desired position. Static pre-load has opposite effects on hard and soft piezoelectric materials. In hard materials, the piezoelectric d_{33} coefficient rises with increasing pre-load. Soft piezoelectric materials show the opposite trend. However, the measurement is more reliable under greater pre-load [76].

Time-Dependent Effect

The measurement result is time dependent when the sample is under pre-load. For all the measurements in [77], the piezoelectric d_{33} coefficient decreases over time, and eventually becomes stable. The specific decay rate is related to the property of the tested piezoelectric material. However, in general, the time to reach stable status for soft materials is longer than that for hard materials [77].

Second-Order Effect

In practical measurements the time-dependent effect can be combined with the frequency-dependent effect. If the frequency sweeps from low frequency to high frequency and then back to low frequency when measuring the piezo-electric d_{33} coefficient, the measured results for a given frequency are different, showing hysteretic behaviour [78]. To eliminate the second-order effect, the measurement can be made after the pre-load has been applied for several hours, to ensure that the piezoelectric d_{33} coefficient has reached a stable region.

Sample and Loading Geometry

When measuring the piezoelectric d_{33} coefficient using the Berlincourt method, the sample under test should be under compression. However, this is not always true in practice. The effect of the sample and loading geometry refers to the shear stress generated when the load contacts the sample and the mechanical interactions between them. To reduce the shear stress, contact electrodes need to be carefully designed. For example, flat electrodes can be used for thin films.

System Calibration

Two calibration points are the zero calibration and gain setting. Normally a non-piezoelectric material is used for zero calibration, and an already tested hard

PZT is used for gain setting. Owing to the frequency-dependent effect, system calibration must be carried out again once the AC force frequency changes.

Environment Effect

In addition to the effects arising from the sample and measurement system, the testing environment also significantly affects the value of the measured piezoelectric d_{33} coefficient. For example, EMI from surrounding electronic instruments disturbs the electric field, so that the measurement is not made under a constant electric field. Also, sudden changes to temperature and humidity can create pyroelectric charges and additional paths for charge leakage. All of these can produce inaccuracy in the measurement.

For ideal solid-state materials, the most simple model for describing deformation behaviour is Hookean linear elasticity. The end point of elastic deformation is fracture. However, the behaviour of many materials in real world is time dependent and non-linear, as discussed earlier, that is related to some combinations of elastic and viscous responses. As to the PVDF, we mainly consider its time- and temperature-dependent responses (especially the temperature, which may decrease d_{33} [81]), which are related to the detection accuracy of the force touch.

References

[1] J. M. Chauvet, E. B. Deschamps, and C. Hillaire, *Dawn of Art: The Chauvet Cave: The Oldest Known Paintings in the World.* New York: Harry N. Abrams, 1996.

[2] J. Aruz and R. Wallenfels, *Art of the First Cities: The Third Millennium BC from the Mediterranean to the Indus.* New York: Metropolitan Museum of Art, 2003.

[3] G. Walker, "A review of technologies for sensing contact location on the surface of a display," *Journal of the Society for Information Display,* vol. 20, Sept., pp. 413–440, 2012.

[4] G. S. Hurst and J. W. C. Colwell, "Discriminating contact sensor," U.S. Patent 3 911 215, Oct. 1975.

[5] J. W. Stetson, "Analog resistive touch panels and sunlight readability," *Information Display,* vol. 22, Dec., pp. 26–30, 2006.

[6] M. H. Ahn, E. S. Cho, and S. J. Kwon, "Effect of the duty ratio on the indium tin oxide (ITO) film deposited by in-line pulsed DC magnetron sputtering method for resistive touch panel," *Applied Surface Science,* vol. 258, Nov., pp. 1242–1248, 2011.

[7] K. Noda and K. Tanimura, "Production of transparent conductive films with inserted SiO_2 anchor layer, and application to a resistive touch panel," *Electronics and Communications in Japan Part II: Electronics,* vol. 84, July, pp. 39–45, 2001.

[8] R. Downs, "Using resistive touch screens for human/machine interface," *Analog Applications Journal Q,* vol. 3, Sept., pp. 5–9, 2005.

[9] G. Barrett and R. Omote, "Projected-capacitive touch technology," *Information Display,* vol. 26, Mar., pp. 16–21, 2010.

[10] E. A. Johnson, "Touch display–a novel input/output device for computers," *Electronics Letters,* vol. 1, Oct., pp. 219–220, 1965.

[11] P.T. Krein and R. D. Meadows, "The electroquasistatics of the capacitive touch panel, *IEEE Transactions on Industry Applications,* vol. 26, May, pp. 529–534, 1990.

[12] S. Hong, J. Yeo, J. Lee, H. Lee, P. Lee, S. S. Lee, and S. H. Ko, "Selective laser direct patterning of silver nanowire percolation network transparent conductor for capacitive touch panel," *Journal of Nanoscience and Nanotechnology,* vol. 15, Mar., pp. 2317–2323, 2015.

[13] T. H. Hwang, W. H. Cui, I. S. Yang, and O. K. Kwons, "A highly area-efficient controller for capacitive touch screen panel systems," *IEEE Transactions on Consumer Electronics,* vol. 56, pp. 1115–1122, May 2010.

[14] K. D. Kim, S. H. Byun, Y. K. Choi, J. H. Baek, H. H. Cho, J. K. Park, and S. W. Kim, "A capacitive touch controller robust to display noise for ultrathin touch screen displays," In 2012 IEEE International Solid-State Circuits Conference Digest of Technical Papers (ISSCC), San Francisco, CA, February 19–23, pp. 116–117.

[15] S. Kim, W. Choi, W. Rim, Y. Chun, H. Shim, H. Kwon, and J. Park, "A highly sensitive capacitive touch sensor integrated on a thin-film-encapsulated active-matrix OLED for ultrathin displays," *IEEE Transactions on Electron Devices*, vol. 58, Oct., pp. 3609–3615, 2011.

[16] R. Adler and P. J. Desmares, "An economical touch panel using SAW absorption," *IEEE Transactions on Ultrasonics, Ferroelectrics, and Frequency Control*, vol. 34, Mar., pp. 195–201, 1987.

[17] K. North and H. D. Souza, "Acoustic pulse recognition enters touch-screen market," *Information Display*, vol. 22, Dec., pp. 22–26, 2006.

[18] M. R. Bhalla and A. V. Bhalla, "Comparative study of various touchscreen technologies," *International Journal of Computer Applications*, vol. 6, Sept., pp. 12–18, 2010.

[19] C. K. Campbell, "Applications of surface acoustic and shallow bulk acoustic wave devices," *Proceedings of the IEEE*, vol. 77, Oct., pp. 1453–1484, 1989.

[20] A. Holzinger, "Finger instead of mouse: Touch screens as a means of enhancing universal access," In *ERCIM Workshop on User Interfaces for All*. Berlin: Springer,pp. 387–397, 2002.

[21] S. Reis, V. Correia, M. Martins, et al., "Touchscreen based on acoustic pulse recognition with piezoelectric polymer sensors," In 2010 IEEE International Symposium on Industrial Electronics (ISIE), pp. 516–520, July 2010.

[22] N. Cohen, "Timeline: a history of touch-screen technology," *National Public Radio*, vol. 26, Dec. 2011.

[23] A. Butler, S. Izadi, and S. Hodges, "SideSight: Multi-touch interaction around small devices," In Proceedings of the 21st Annual ACM Symposium on User Interface Software and Technology, pp. 201–204, Oct. 2008.

[24] R.D. Blanchard, "Infrared touch panel with improved sunlight rejection," U.S. Patent 6 677 934. Jan. 13, 2004.

[25] T.E. Masters, R.W. Knetsch, H.A. Grice, Jr., and J. Deacon, "Apparatus and method to improve resolution of infrared touch systems," U.S. Patent 6 429 857, Aug. 6, 2002.

[26] M.L. Smith, "Battery-operated data collection apparatus having an infra-red touch screen data entry device," U.S. Patent 4 928 094, May 22, 1990.

[27] O. Raymudo, "Iphone 6S display teardown reveals how 3D touch sensors actually work," Message posted to Macworl, Oct. 2015.

[28] S. Gao, V. Arcos and A. Nathan, "Piezoelectric vs. capacitive based force sensing in capacitive touch panels,"1 *IEEE Access*, vol. 4, pp. 3769–3774, 2016.

[29] S. Gao and A. Nathan, "P-180: Force Sensing Technique for Capacitive Touch Panel," *SID Symposium Digest of Technical Papers*, vol. 47, no. 1, May, pp. 1814–1817, 2016.

[30] A. Nathan and S. Gao, "Interactive displays: The next omnipresent technology [Point of View]," *Proceedings of the IEEE*, vol. 104, Aug., pp. 1503–1507, 2016.

[31] S. Jeon, S.E. Ahn, L. Song, et al., "Gated three-terminal device architecture to eliminate persistent photoconductivity in oxide semiconductor photosensor arrays," *Nature Materials*, vol. 11, Apr., pp. 301–305, 2012.

[32] S.E. Ahn, I. Song, S. Jeon, et al., "Metal oxide thin film phototransistor for remote touch interactive displays," *Advanced Materials*, vol. 24, May, pp. 2631–2636, 2012.

[33] S. Jeon, S.E. Ahn, I. Song, et al., "Dual gate photo-thin film transistor with high photoconductive gain for high reliability, and low noise flat panel transparent imager," In Electron Devices Meeting (IEDM), pp. 331–334, Dec. 2011.

[34] S. Jeon, S. Park, I. Song, et al., "Nanometer-scale oxide thin film transistor with potential for high-density image sensor applications," *ACS Applied Materials & Interfaces*, vol. 3, Dec., pp. 1–6, 2010.

[35] S. Lee, S. Jeon, R. Chaji, and A. Nathan, "Transparent semiconducting oxide technology for touch free interactive flexible displays," *Proceedings of the IEEE*, Vol. 103, Apr., pp. 644–664, 2015.

[36] Z. Hua and W.L. Ng, "Speech recognition interface design for in-vehicle system," In *Proceedings of the 2nd International Conference on Automotive User Interfaces and Interactive Vehicular Applications*, ACM, Nov., pp. 29–33, 2010.

[37] M.F. Ruzaij, S. Neubert, N. Stoll, and K. Thurow, "Hybrid voice controller for intelligent wheelchair and rehabilitation robot using voice recognition and embedded technologies," *Journal of Advanced Computational Intelligence and Intelligent Informatics*, vol. 20, July, pp. 615–622, 2016.

[38] S. Li, X. Zhang, F.J. Kim, R.D. da Silva, D. Gustafson, and W. R. Molina, "Attention-aware robotic laparoscope based on fuzzy interpretation of eye-gaze patterns," *Journal of Medical Devices*, vol. 9, Dec., 041007, 2015.

[39] K. Zinchenko, C. Y. Wu, and K. T. Song, "A study on speech recognition control for a surgical robot," *IEEE Transactions on Industrial Informatics*, vol. 13, Apr., pp. 607–615, 2017.

[40] H.M. Do, W. Sheng, and M. Liu, "Human-assisted sound event recognition for home service robots," *Robotics and Biomimetics*, vol. 3, issue 7, Dec., 2016.

[41] O.K. Kwon, J.S. An, and S.K. Hong, "Capacitive touch systems with styli for touch sensors: A review", *IEEE Sensors Journal*, vol. 18, issue 12, pp. 4832–4846, 2018.

[42] S. Gao, J. Lai, and A. Nathan, "Reduction of common-mode noise in capacitive touch panels by correlated double sampling," *IEEE/OSA Journal of Display Technology*, vol. 12, no. 6, pp. 639–645, 2016.

[43] S. Gao, J. Lai, and A. Nathan, "Reduction of noise spikes in touch screen systems by low pass spatial filtering," *IEEE/OSA Journal of Display Technology*, vol. 12, no. 9, pp. 957–963, 2016.

[44] 3 M Company, "Touch technology brief: Projected capacitive technology", 2011.

[45] G. Walker, "Touch and the Apple iPhone," Veritas et Visus, www.veritasetvisus.com/touch_panel.htm (accessed 1 Apr. 2018).

[46] G. Walker, "Fundamentals of touch technologies," In Sunday Short Course (S-4), SID Display Week 2013.

[47] S. Gao, J. Lai, C. Micou, and A. Nathan, "Reduction of common mode noise and global multivalued offset in touch screen systems by correlated double sampling," *Journal of Display Technology*, vol. 12, June, pp. 639–645, 2016.

[48] H. Shin, S. Ko, H. Jang, I. Yun, and K. Lee, "A 55 dB SNR with 240 Hz frame scan rate mutual capacitor 30× 24 touch-screen panel read-out IC using code-division multiple sensing technique," In Solid-State Circuits Conference Digest of Technical Papers (ISSCC), Feb., pp. 388–389, 2013.

[49] N. Miura, S. Dosho, S. Takaya, et al., "12.4 A 1 mm-pitch 80× 80-channel 322 Hz-frame-rate touch sensor with two-step dual-mode capacitance scan," In Solid-State Circuits Conference Digest of Technical Papers (ISSCC), 2014 IEEE International, Feb. pp. 216–217, 2014.

[50] J.H. Yang, S.H. Park, J.M. Choi, et al., "A highly noise-immune touch controller using filtered-delta-integration and a charge-interpolation technique for 10.1-inch capacitive touch-screen panels," In Solid-State Circuits Conference Digest of Technical Papers (ISSCC), 2013 IEEE International. IEEE, Feb., pp. 390–391, 2013.

[51] H. Jang, H. Shin, S. Ko, I. Yun, and K. Lee, "12.5 2D Coded-aperture-based ultra-compact capacitive touch-screen controller with 40 reconfigurable

channels," In Solid-State Circuits Conference Digest of Technical Papers (ISSCC), 2014 IEEE International, Feb., pp. 218–219, 2014.

[52] S. Ko, H. Shin, J. Lee, et al., "Low noise capacitive sensor for multi-touch mobile handset's applications," In Solid State Circuits Conference (A-SSCC), 2010 IEEE Asian, Nov. pp. 1–4, 2010.

[53] S. Ko, H. Shin, H. Jang, I. Yun, and K. Lee, "A 70 dB SNR capacitive touch screen panel readout IC using capacitor-less trans-impedance amplifier and coded Orthogonal Frequency-Division Multiple Sensing scheme," In VLSI Circuits (VLSIC), 2013 Symposium on, June, pp. C216–C217. IEEE, 2013.

[54] J.E. Park, D.H. Lim, and D.K. Jeong, "A reconfigurable 40-to-67 dB SNR, 50-to-6400 Hz frame-rate, column-parallel readout IC for capacitive touch-screen panels, " *IEEE Journal of Solid-State Circuits*, vol. 49, Oct., pp. 2305–2318, 2014.

[55] H.E. Jeong, J.K. Lee, H.N. Kim, S.H. Moon, and K.Y. Suh, "A nontransferring dry adhesive with hierarchical polymer nanohairs," *Proceedings of the National Academy of Sciences*, vol. 106, Apr., pp. 5639–5644, 2009.

[56] D.S. Hecht, D. Thomas, L. Hu, et al., "Carbon-nanotube film on plastic as transparent electrode for resistive touch screens," *Journal of the Society for Information Display*, vol. 17, Nov., pp. 941–946, 2009.

[57] K.L. Du, and M.N. Swamy, *Wireless Communication Systems: From RF Subsystems to 4 G Enabling Technologies*. Cambridge: Cambridge University Press, 2010.

[58] Course Note of Hyper Physics, Department of Physics and Astronomy, Georgia State University, http://hyperphysics.phy-astr.gsu.edu/hbase/elec tric/capchg.html (accessed 3 May 2018).

[59] M. Xu, J. Sun, and F.C. Lee, "Voltage divider and its application in the two-stage power architecture," In Applied Power Electronics Conference and Exposition, APEC'06. Twenty-First Annual IEEE, Mar., pp. 499–505, 2006.

[60] H. Philipp, "Charge transfer capacitance measurement circuit," U.S. Patent 6 466 036, Oct. 2002.

[61] S.Y. Peng, M.S. Qureshi, P.E. Hasler, A. Basu, and F.L. Degertekin, "A charge-based low-power high-SNR capacitive sensing interface circuit," *IEEE Transactions on Circuits and Systems I*, vol. 55, Aug., pp. 1863–1872, 2008.

[62] S. Gao, D. McLean, J. Lai, C. Micou, and A. Nathan, "Reduction of noise spikes in touch screen systems by low pass spatial filtering," *Journal of Display Technology*, vol. 12, Sept., pp. 957–963, 2016.

[63] S. Gao, J. Lai, and A. Nathan, "Fast readout and low power consumption in capacitive touch screen panel by downsampling," *Journal of Display Technology*, vol. 12, Nov., pp. 1417–1422, 2016.

[64] N.G. Sevastopoulos and D.A. LaPorte, "Linear Technology Corporation, Flexible monolithic continuous-time analog low-pass filter with minimal circuitry," U.S. Patent 6 344 773, Feb. 2002.

[65] H. Akhtar and R. Kakarala, "A methodology for evaluating accuracy of capacitive touch sensing grid patterns," *Journal of Display Technology*, vol. 10, Aug., pp. 672–682,2014.

[66] H. Akhtar and R. Kakarala, "A comparative analysis of capacitive touch panel grid designs and interpolation methods," In 2014 IEEE International Conference on Image Processing (ICIP), Oct., pp. 5796–5800, 2014.

[67] C. Luo, M.A. Borkar, A.J. Redfern, and J.H. McClellan, "Compressive sensing for sparse touch detection on capacitive touch screens," *IEEE Journal on Emerging and Selected Topics in Circuits and Systems*, vol. 2, Sept., pp. 639–648, 2012.

[68] H. Kawai, "The piezoelectricity of poly (vinylidene fluoride)," *Japanese Journal of Applied Physics*, vol. 8, July, pp. 975–976, 1969.

[69] F. Liu, N.A. Hashim, Y. Liu, M.M. Abed, and K. Li, "Progress in the production and modification of PVDF membranes," *Journal of Membrane Science*, vol. 375, June, pp. 1–27, 2011.

[70] T. Goldacker, V. Abetz, R. Stadler, I. Erukhimovich, and L. Leibler, "Non-centrosymmetric superlattices in block copolymer blends," *Nature*, vol. 398, Mar., pp. 137–139, 1999.

[71] J. Briscoe, N. Jalali, P. Woolliams, et al., "Measurement techniques for piezoelectric nanogenerators," *Energy & Environmental Science*, vol. 6, pp. 3035–3045, 2013.

[72] M.G. Cain, ed., *Characterisation of Ferroelectric Bulk Materials and Thin Films*, vol. 2. Dordrecht, the Netherlands: Springer, 2014.

[73] A. Nathan and B. Henry, *Microtransducer CAD: Physical and Computational Aspects*, Vienna, Springer, 1999.

[74] D. Berlincourt, H. Jaffe, and L.R. Shiozawa, "Electroelastic properties of the sulfides, selenides, and tellurides of zinc and cadmium," *Physical Review*, vol. 129, Feb., pp. 1009–1017, 1963.

[75] D.A. Hall, "Review nonlinearity in piezoelectric ceramics," *Journal of Materials Science*, vol. 36, Oct., pp. 4575–4601, 2001.

[76] J. Wooldridge, A. Muniz-Piniella, M. Stewart, T.A.V. Shean, P.M. Weaver, and M.G. Cain, "Vertical comb drive actuator for the measurement of piezo-electric coefficients in small-scale systems," *Journal of Micromechanics and Microengineering*, vol. 23, Feb., 035028, 2013.

[77] M.G. Cain, M. Stewart, and M.G. Gee, "Degradation of piezoelectric materials," Teddington: National Physical Laboratory, Jan. 1999.

[78] J.F. Blackburn and M.G. Cain, "Nonlinear piezoelectric resonance: A theoretically rigorous approach to constant I– V measurements," *Journal of Applied Physics*, vol. 100, Dec., 114101, 2006.

[79] G.S. Hurst and J.W.C. Colwell, "Elographics Inc, Discriminating contact sensor," U.S. Patent 3 911 215, Oct. 1975.

[80] 740 PVDF Material Data Sheet, Kynar Corp., www.professionalplastics.com /professionalplastics/Kynar740DataSheet.pdf (accessed 25 May 2018).

[81] G.D. Jones, R.A. Assink, T.R. Dargaville, et al., "Characterization, performance and optimization of PVDF as a piezoelectric film for advanced space mirror concepts, (No. SAND2005-6846)," Sandia National Laboratories, Nov. 2005.

[82] "Interfacing Piezo Film to Electronics," Measurement Specialties Inc., Application Note 01800004–000, March 2006.

[83] J. Karki, "Signal Conditioning Piezoelectric Sensors," Application Report – SLOA033A, Texas Instruments, 2000.

[84] M. Kang, J. Kim, B. Jang, Y. Chae, J.H. Kim, and J.H. Ahn, "Graphene-based three-dimensional capacitive touch sensor for wearable electronics," ACS Nano, vol. 11, July, pp. 7950–7957, 2017.

[85] Y. Zi, L. Lin, J. Wang, et al., "Triboelectric–pyroelectric–piezoelectric hybrid cell for high-efficiency energy-harvesting and self-powered sensing," *Advanced Materials*, vol. 27(14), Apr., pp. 2340–2347, 2015.

[86] S. Filiz, B.Q. Huppi, K. Wang, P.W. Richards, and G.A.R.G. Vikram, "Apple Inc, Force detection in touch devices using piezoelectric sensors," U.S. Patent 9 983 715, May 2018.

[87] Y.L. Chen, S. Wang, Y. Shimizu, S. Ito, W. Gao, and B.F. Ju, "An in-process measurement method for repair of defective microstructures by using a fast tool servo with a force sensor," *Precision Engineering*, vol. 39, Jan., pp.134–142, 2015.

[88] P. Wang, Y. Fu, B. Yu, Y. Zhao, L. Xing, and X. Xue, "Realizing room-temperature self-powered ethanol sensing of ZnO nanowire arrays by combining their piezoelectric, photoelectric and gas sensing characteristics," *Journal of Materials Chemistry A*, vol. 3, pp. 3529–3535, 2015.

[89] E. Aranda-Michel, J. Yi, J. Wirekoh, et al., "Miniaturized robotic end-effector with piezoelectric actuation and fiber optic sensing for minimally invasive cardiac procedures," *IEEE Sensors Journal*, vol. 18, June, pp. 4961–4968, 2018.

[90] A. Spanu, L. Pinna, F. Viola, et al., "A high-sensitivity tactile sensor based on piezoelectric polymer PVDF coupled to an ultra-low voltage organic transistor," *Organic Electronics*, vol. 36, Sept., pp. 57–60, 2016.

[91] C. Dagdeviren, Y. Su, P. Joe, et al., "Conformable amplified lead zirconate titanate sensors with enhanced piezoelectric response for cutaneous pressure monitoring," *Nature Communications*, Vol. 5, Aug., 4496, 2014.

[92] "Specialty glass products technical reference document," Corning Inc., Aug. 2012, https://abrisatechnologies.com/specs/Corning%200211% 20Microsheet%20Spec%20Sheet%2012_10.pdf (accessed 17 April 2019).

[93] B. Mohammadi, A.A. Yousefi, and S.M. Bellah, "Effect of tensile strain rate and elongation on crystalline structure and piezoelectric properties of PVDF thin films," *Polymer Testing*, Vol. 26, Feb., pp. 42–50, 2007.

[94] J. G. Speight, "Norbert Adolph Lange," In *Lange's Handbook of Chemistry* (16 ed.). New York: McGraw-Hill, 2005.

[95] "The future is flexible: Corning willow glass," Technique Datasheet, Corning Inc., 2012.

[96] B. Ren and C.J. Lissenden, "Pvdf multielement lamb wave sensor for structural health monitoring," *IEEE Transactions on Ultrasonics, Ferroelectrics, and Frequency Control*, vol. 63, Jan., pp. 178–185, 2016.

[97] M. Crescentini, M. Bennati, M. Carminati, and M. Tartagni, "Noise limits of CMOS current interfaces for biosensors: A review," *IEEE Transactions on Biomedical Circuits and Systems*, vol. 8, Apr., pp. 278–292, 2014.

[98] G. Walker, "Fundamentals of Projected-Capacitive Touch Technology," Society for Information Display Display Week, 2014, http://walkermobile .com/Touch_technologies_Tutorial_Latest_Version.pdf (Accessed 17 April 2019).

[99] Properties of poled PVDF, Acoustics Inc., www.acoustics.co.uk/pal/wp-content/uploads/2015/11/Properties-of-poled-PVDF.pdf (accessed 5 May 2018).

[100] X. Wu, G. Zhong, L. D'Arsié, et al., "Growth of continuous monolayer graphene with millimeter-sized domains using industrially safe conditions, " *Scientific Reports*, vol. 6, Feb., 21152, 2016.

[101] R.R. Nair, P. Blake, A.N. Grigorenko, et al., "Fine structure constant defines visual transparency of grapheme," *Science*, vol. 320, June, pp. 1308–1308, 2008.

[102] S.K. Lee, B.J. Kim, H. Jang, et al., "Stretchable graphene transistors with printed dielectrics and gate electrodes," Nano Letters, vol. 11, Oct., pp. 4642–4646, 2011.

[103] Z. Liu, Q. Liu, Y. Huang, et al., "Organic photovoltaic devices based on a novel acceptor material: Grapheme," *Advanced Materials*, vol. 20, Oct., pp. 3924–3930, 2008.

[104] T.H. Han, Y. Lee, M.R. Choi, et al., "Extremely efficient flexible organic light-emitting diodes with modified graphene anode," *Nature Photonics*, vol. 6, Feb., pp. 105–110, 2012.

[105] R.H. Kim, M.H. Bae, D.G. Kim, et al., "Stretchable, transparent graphene interconnects for arrays of microscale inorganic light emitting diodes on rubber substrates," *Nano Letters*, Vol. 11, Aug., pp. 3881–3886, 2011.

[106] J.A. Rogers, "Electronic materials: Making graphene for macroelectronics," *Nature Nanotechnology*, Vol. 3, May, pp. 254–255, 2008.

[107] F Bonaccorso, Z. Sun, T. Hasan, and A.C. Ferrari, "Graphene photonics and optoelectronics," *Nature Photonics*, vol. 4, Sept., pp. 611–622, 2010.

[108] S.H. Bae, Y. Lee, B.K. Sharma, H.J. Lee, J.H. Kim, and J.H. Ahn, "Graphene-based transparent strain sensor," *Carbon*, vol. 51, Jan., pp. 236–242, 2013.

[109] C. Lee, X. Wei, J.W. Kysar, and J. Hone, "Measurement of the elastic properties and intrinsic strength of monolayer grapheme," *Science*, vol. 321, July, pp. 385–388, 2008.

[110] K.S. Novoselov, V.I. Fal, L. Colombo, P.R. Gellert, M.G. Schwab, and K. Kim, "A roadmap for graphene," *Nature*, vol. 490, Oct., pp. 192–200, 2012.

[111] www.mit.edu/~6.777/matprops/ito.htm (accessed 30 Sept. 2017).

[112] X. Li, W. Cai, J. An, et al., "Large-area synthesis of high-quality and uniform graphene films on copper foils," *Science*, vol. 324, pp. 1312–1314, 2009.

[113] X. Liang, B.A. Sperling, and I. Calizo, "Toward clean and crackless transfer of graphene," *ACS Nano*, vol. 5, issue 11, pp. 9144–9153, 2011.

[114] S. Cha, S.M. Kim, H. Kim, et al., "Porous PVDF as effective sonic wave driven nanogenerators," *Nano Letters*, vol. 11, Nov., pp. 5142–5147, 2011.

[115] C. Chang, V.H. Tran, J. Wang, Y.K. Fuh, and L. Lin, "Direct-write piezoelectric polymeric nanogenerator with high energy conversion efficiency," *Nano Letters*, vol. 10, Jan., pp. 726–731, 2010.

[116] S.W. Choi, S.M. Jo, W.S. Lee, and Y.R. Kim, "An electrospun poly (vinylidene fluoride) nanofibrous membrane and its battery applications," *Advanced Materials*, Vol. 15, Dec., pp. 2027–2032, 2003.

[117] T. Furukawa, "Ferroelectric properties of vinylidene fluoride copolymers," *Phase Transitions*, vol. 18, Aug., pp. 143–211, 1989.

[118] S. Fujisaki, H. Ishiwara, and Y. Fujisaki, "Low-voltage operation of ferroelectric poly (vinylidene fluoride-trifluoroethylene) copolymer capacitors and metal-ferroelectric-insulator-semiconductor diodes," *Applied Physics Letters*, vol. 90, Apr., 162902, 2007.

[119] R.G. Kepler and R.A. Anderson, "Ferroelectric polymers," *Advances in Physics*, vol. 41, pp. 1–57, 1992.

[120] Y. Xu, *Ferroelectric Materials and Their Applications*. Philadelphia: Elsevier, 2013.

[121] X. He and K. Yao, "Crystallization mechanism and piezoelectric properties of solution-derived ferroelectric poly (vinylidene fluoride) thin films," *Applied Physics Letters*, vol. 89, Sept., 112909, 2006.

[122] H.A. Sodano, D.J. Inman, and G. Park, "A review of power harvesting from vibration using piezoelectric materials," *Shock and Vibration Digest*, vol. 36, May, pp. 197–206, 2004.

[123] A. Carroll and G. Heiser, "An analysis of power consumption in a smartphone," *USENIX Annual Technical Conference*, vol. 14, June, pp. 21–21, 2010.

[124] M. Mehendale, S. Das, M. Sharma, et al., "A true multistandard, programmable, low-power, full HD video-codec engine for smartphone SoC," In Solid-State Circuits Conference Digest of Technical Papers (ISSCC), 2012 IEEE International, Feb., pp. 226–228, 2012.

[125] A. Sampson, W. Dietl, E. Fortuna, D. Gnanapragasam, L. Ceze, and D. Grossman, "EnerJ: Approximate data types for safe and general low-power computation," *ACM SIGPLAN Notices*, vol. 46, June, pp. 164–174, 2011.

[126] C. Gomez, J. Oller, and J. Paradells, "Overview and evaluation of bluetooth low energy: An emerging low-power wireless technology," *Sensors*, vol. 12, Aug., pp. 11734–11753, 2012.

[127] E. Cuervo, A. Balasubramanian, D.K. Cho, et al., "MAUI: making smartphones last longer with code offload," In Proceedings of the 8th International Conference on Mobile Systems, Applications, and Services, June, pp. 49–62, 2010.

[128] D. Li and W.G. Halfond, "An investigation into energy-saving programming practices for android smartphone app development," In Proceedings of the 3rd International Workshop on Green and Sustainable Software, June, pp. 46–53, 2014.

[129] P.M.Y. Fan, O.Y. Wong, M.J. Chung, T.Y. Su, X. Zhang and P.H. Chen, "Energy harvesting techniques: Energy sources, power management and conversion," In 2015 European Conference on Circuit Theory and Design (ECCTD), Aug., pp. 1–4, 2015.

Cambridge Elements ☰

Flexible and Large-Area Electronics

Ravinder Dahiya

University of Glasgow

Ravinder Dahiya is Professor of Electronic and Nanoengineering and an EPSRC
Fellow at the University of Glasgow. He is the Director of Electronics Systems Design Centre
at the University of Glasgow and leads the multidisciplinary group, Bendable Electronics
and Sensing Technologies (BEST). He is President-Elect and a Distinguished Lecturer of the
IEEE Sensors Council and serves on the Editorial Boards of the *Scientific Reports, IEEE Sensors
Journal,* and *IEEE Transactions on Robotics.* He is an expert in the field of flexible
and bendable electronics, robotics and electronic skin.

Luigi G. Occhipinti

University of Cambridge

Luigi G. Occhipinti is Director of Research at the University of Cambridge, Engineering
Department, and Deputy Director and COO of the Cambridge Graphene
Centre. He is the founder and CEO of Cambridge Innovation Technologies Consulting
Limited, providing research and innovation within both the health care and medical
fields. He is a recognized expert in printed, organic, and large-area electronics
and integrated smart systems with more than twenty years' experience in the
semiconductor industry, and is a former R&D Senior Group Manager and Programs
Director at STMicroelectronics.

About the series

This innovative series provides authoritative coverage of the state-of-the-art in bendable
and large-area electronics. Specific Elements provide in-depth coverage of key
technologies, materials and techniques for the design and manufacturing of
flexible electronic circuits and systems, as well as cutting-edge insights into emerging
real-world applications. This series is a dynamic reference resource for graduate
students, researchers, and practitioners in electrical engineering, physics, chemistry
and materials science.

Cambridge Elements $^{\equiv}$

Flexible and Large-Area Electronics

Elements in the series

Bioresorbable Materials and Their Application in Electronics
Xian Huang

Organic and Amorphous-Metal-Oxide Flexible Analogue Electronics
Vincenzo Pecunia et al.

Large-Area Electronics Based on Micro/Nanostructures and the Manufacturing Technologies
Carlos García Núñez et al.

A Flexible Multi-Functional Touch Panel for Multi-Dimensional Sensing in Interactive Displays
Shuo Gao and Arokia Nathan

A full series listing is available at www.cambridge.org/eflex

Printed in the United States
By Bookmasters